Relativity: A Journey Through Warped Space
and Time

Daniel R. Mayerson · Anthony M. Charles ·
Joseph E. Golec

Relativity: A Journey Through Warped Space and Time

 Springer

Daniel R. Mayerson
Institut de Physique Théorique
Université Paris Saclay
Gif-sur-Yvette, France

Anthony M. Charles
Institute for Theoretical Physics
KU Leuven
Leuven, Belgium

Joseph E. Golec
Randall Laboratory
University of Michigan
Ann Arbor, MI, USA

ISBN 978-3-030-18913-6 ISBN 978-3-030-18914-3 (eBook)
https://doi.org/10.1007/978-3-030-18914-3

This Springer imprint is published by the registered company Springer Nature Switzerland AG.
The registered company address is: Gewerbestrasse 11, 6330 Cham, Switzerland

Preface

If you've managed to stumble upon these notes, either as an educator or a student, odds are that you've heard of relativity, Einstein's famous theory of space, time, and gravitation. It is arguably his most celebrated accomplishment in physics, and it is certainly one of the most important ideas to emerge in physics in the twentieth century. Broadly speaking, relativity unifies space and time into a single entity, the geometry of which is determined by the gravitational fields emanating from massive objects in the universe.

There are plenty of popular science books and instructional videos that will give you this brief description of relativity, perhaps also telling you about some of the amazing consequences of the theory, such as time dilation or black holes. Here, we want to go *further*, by diving into the nuts and bolts of relativity and developing a deep understanding of these phenomena and how they come about. Our goal is to satisfy any thirst the reader may have for the details and inner workings of relativity, a thirst you may have after reading or seeing such popular scientific books or videos. We want to do justice to the remarkable simplicity and structure of the concepts and physical insights that lie at the heart of relativity. To accomplish this, we will motivate relativity from first principles while simultaneously building up the mathematical framework needed to understand the nuts and bolts of how relativity actually works.

These notes are introductory and aimed at students not necessarily fluent with advanced mathematical and physical concepts, such as researchers in other fields, beginning undergraduate students, or even high school students or other interested people. Science enthusiasts who discover and read about relativity through popular science and get excited about the subject unfortunately have very few resources available to them to learn more. Most textbooks or courses are aimed at (advanced) undergraduate students or even graduate students and assume much more mathematical background than the typical non-expert science enthusiast possesses. These notes give a pedagogical treatment of the subject and demonstrate that an advanced mathematical background is not necessary to understand relativity. In fact, relativity is based on some simple and elegant (but also profound!) notions that can naturally be understood through simple thought experiments. Once these concepts are clear, the mathematical underpinnings of these ideas become natural and straightforward to learn and apply. We hope these notes help dispel the common myth that relativity is inaccessible and too esoteric for the average person to comprehend.

Moreover, we want to emphasize that relativity is not an abstract mathematical theory with no practical applications. For example, relativistic calculations are required to make a wide range of modern technology work, including GPS satellites and semiconductors. The theory of relativity forms the backbone of a wide variety of modern physics research disciplines, including astrophysics, cosmology, and even string theory. Relativity also remains an active and exciting area of research, even though the theory itself is over one hundred years old: The recent and first-ever direct detection of gravitational waves by LIGO resulted in the 2017 Nobel Prize in physics. By learning the details and getting hands-on experience with calculations in relativity, students will be able to achieve a deeper understanding of these areas of study and be more suited to understand present and future developments.

Another important reason to study relativity is to learn how to think like a scientist, i.e., to learn how to think critically about our world. Einstein developed his theory by first identifying inconsistencies in the conventionally agreed-upon laws of physics at the time and then coming up with a new set of laws that would fix these inconsistencies. By studying the precise details of this process, students will learn how to think critically about scientific theories, improve existing ideas, and come up with experiments that challenge conventional ways of thinking. These skills are indispensable not only to be a good scientist, but also to be a well-informed and thoughtful citizen of the world.

We could go on listing dozens more ways that learning a subject like relativity is beneficial for students, but the last one we will mention here is simply that relativity is an amazing, elegant, and beautiful description of our universe, and we hope that anyone who takes time to learn its intricacies will come out with a vigor and passion for understanding the mysteries of our universe.

Gif-sur-Yvette, France   Daniel R. Mayerson
Leuven, Belgium   Anthony M. Charles
Ann Arbor, USA   Joseph E. Golec

Acknowledgements

First and foremost, we want to thank the bright, enthusiastic high school students that participated in our MMSS course in 2017 and 2018, for making the entire teaching experience exciting and enjoyable for us. You are the future of physics! We also want to thank the organizers of the MMSS program in the Department of Mathematics at the University of Michigan for the organization of the MMSS program.

We wish to thank Henriette Elvang and Ratindranath Akhoury for valuable feedback, comments, and suggestions in preparing these notes, especially for the online version.

When employed at the University of Michigan, D.R.M. and A.M.C. were supported in the Department of Physics by the US Department of Energy under grants DE-FG02-95ER4089 and DE-SC0007859. D.R.M. is currently supported by the ERC Starting Grant 679278 Emergent-BH. A.M.C. is currently supported in part by the KU Leuven C1 Grant ZKD1118 C16/16/005, the National Science Foundation of Belgium (FWO) Grant G.001.12 Odysseus, and by the European Research Council Grant No. ERC-2013-CoG 616732 HoloQosmos. These course notes were originally created as part of the MMSS program in 2017 and 2018.

Contents

Chapter 1
Introduction

Welcome to the wonderful world of relativity! These course notes are at an introductory level, aimed at an enthusiastic student or learner who does not necessarily have advanced mathematical or physical knowledge past the high school level. These notes are also intended to serve as a guideline for any educators aspiring to teach an expository course on relativity. Either way, we hope that you will enjoy your time exploring space and time with us!

1.1 Lecture Note Structure and Legend

We should first point out that these notes are very long, and it may seem daunting to start reading or studying them. The length should not intimidate anybody, though. The goal of these notes is to cover many different topics that a reader might be interested in, so there is much more content here than the average reader will need in order to grasp the fundamentals of relativity.

Barred parts of these notes, i.e. sections with a *grey bar* to the left of it, are extra material and are not essential. In a first reading this material can all safely be skipped. Advanced students (or those with lots of time) may want to study this extra material for a deeper insight into the material. Note that barred sections later in the notes may depend on barred material earlier on.

There are some exercises in the text and some at the end of sections or chapters. They are there to help you process and familiarize yourself with the material presented. Some exercises may ask you to fill in gaps in the notes or think about a consequence of a concept that was just introduced.

The approximate *difficulty* of an exercise is marked with a number of stars. An exercise with no stars should not be too difficult; those marked with * are more difficult; those marked with ** are very difficult; and those marked with *** would

© Springer Nature Switzerland AG 2019
D. R. Mayerson et al., *Relativity: A Journey Through Warped Space and Time*,
https://doi.org/10.1007/978-3-030-18914-3_1

probably be considered a challenging problem for graduate students taking a graduate level course on relativity—attempt at your own peril!

The grey boxes marked with a *Discussion* title are meant to be motivational discussion questions to introduce concepts; ideally, they will be tackled in small groups of students. Usually, the concepts in these discussion questions are then elaborated on and discussed in the text following the grey box of the discussion.

Appendix A is a quick reference section that succinctly summarizes all of the major formulas and equations found throughout the notes. Students may find it useful to keep a copy of this section with them as they go through the notes in order to double-check any equations they might use to solve exercises.

Appendix B contains solutions and hints for some of the exercises and discussion questions. For exercises that ask for numerical results, the solutions are present so that students can check their final answers, but details of how to obtain the solutions may not be included. For discussion questions, a brief summary of the correct answer is often given.

1.2   Notes for Educators

These notes were originally created to be supplemental to the two week long course *Relativity: A Journey through Warped Space and Time* that taught special and general relativity as part of the MMSS (Michigan Math and Science Scholars) summer program for high school students at the University of Michigan.[1] If you are an educator or instructor and are interested in using these course notes as a guide or inspiration for your own introductory special and/or general relativity course for beginning undergraduates, researchers from unrelated fields, or even high school students—that's great! Depending on the level of the students and the length of your course, you may want to skip the extra barred material and possibly other sections as well. The background (especially relating to math) of the students we taught in the MMSS course using these notes varied, and different students were able to process different amounts of material and exercises accordingly.

Relativity is not usually taught at a very basic level (such as beginning undergraduate or high school level) as it is often assumed that the level and amount of mathematical tools needed is well beyond what they know. While this may be true for most existing relativity courses, we wish to stress that it is certainly possible to teach relativity at this level. Throughout our course, we would first introduce new concepts using the relevant *Discussion* boxes featured throughout the notes in a group discussion in class. Only once the students were familiar with the concept intuitively did we introduce the relevant mathematics. In this way, seeing the actual mathematical machinery is almost an afterthought and not the crucial part of the learning experience!

---

[1] See the MMSS web-site here: http://www.math.lsa.umich.edu/mmss/.

In particular, special relativity does not require the introduction of many math concepts beyond the Lorentz transformations in order for students to be able to dive into exciting qualitative, quantitative, conceptual, and calculational aspects of the theory. Conventional approaches to teaching general relativity, on the other hand, rely heavily on first providing the students with a number of tools from differential geometry such as manifolds and tensors. We found it rather unwieldy and also unnecessary to introduce these very formal mathematical details. Instead, we had the students gain an intuitive understanding of concepts like curvature, parallel transport, geodesics, etc., via thought experiments involving lines, paths, and imaginary ants on curved spaces (such as spheres)—see the *Discussion* boxes in Chap. 4. By the end of the course, we did not expect students to be able to work out all mathematical details of general relativity, but they *would* be able to explain the underlying concepts. For example, they wouldn't necessarily be able to work with the geodesic equation, but they understood what a geodesic was and could explain what traveling on a geodesic meant for an observer moving on a sphere.

To give a guideline as to the timeline of our course, we provide in Table 1.1 a rough schedule of the material presented during the two week course. Each of the ten days consisted of roughly six hours of class time (except the last day, which was a half day). In addition to the "regular" class activities listed below, there was also a visit to the local planetarium, as well as two afternoons (i.e. the afternoons of Monday and Tuesday in the second week) of computer labs introducing Mathematica and performing simple exercises related to curvature and orbits in the Schwarzschild solution using a pre-made Mathematica notebook. On both Thursday afternoons, there was an additional "movie reel" activity where science-fiction movie clips were analyzed for accuracy in the context of special (first Thursday) and general (second Thursday) relativity.

In teaching the students about curvature, parallel transport, and geodesics (discussion questions 4.D, 4.E, 4.F), it was extremely useful for the students to have actual pieces of papers and beach balls on which to draw lines and geodesics, so that they could visualize the relevant concepts more easily. When we started discussing the subjects of curvature, parallel transport, and geodesics, we first had the students think about discussion questions 4.D, 4.E, and 4.F. Afterward, we introduced the relevant mathematical formulas (e.g. the Christoffel symbols given in (4.47)) with some simple examples (e.g. flat space). Then, we had them work through the simple 1D and 2D examples given in Exercise 4.30–4.32, which the students found helpful to cement their understanding of the mathematical details.

Importantly, there is much more content in these notes than can feasibly be taught in great detail within such a two-week course. We encourage any educators using these notes to take our schedule as an inspiration, while taking time to focus on the topics that the students are most interested in.

Table 1.1 A sample of the schedule used when teaching a two-week relativity course based on these notes

Week	Monday	Tuesday	Wednesday	Thursday	Friday
1	Introduction,	Special relativity (3.1–3.2),	Paradoxes (cont.) (3.2.5),	Relativistic mechanics (3.4),	Equivalence principles (4.1),
	Math fundamentals (2.1),	Paradoxes (3.2.5)	Spacetime (3.3)	Wrap-up and review	Intro to curvature (4.2.1),
	Physics fundamentals (2.2)				Curvilinear coordinates (2.1.6)
2	Parallel transport, geodesics (4.2.4–4.2.5),	Riemann tensor and Ricci tensor (4.4.2–4.4.4),	Schwarzschild solution (cont.) (4.6.1),	Cosmology (4.6.4),	Time machines (4.6.6),
	Straight lines in different coordinates (4.2.2),	Einstein equation (4.5),	Black holes (4.6.2),	Gravitational waves (4.6.5),	Beyond general relativity (4.7)
	Covariant derivatives (4.2.3)	Schwarzschild solution (4.6.1)	Wormholes (4.6.3)	Extra topics	

1.3   Notes for Students

Perhaps you are unfamiliar with relativity and would like to use these notes to learn more about special and general relativity at an introductory level. Or, you might already be knowledgeable about aspects of relativity, but you want to sharpen up your understanding of more in-depth topics, like black holes and gravitational waves. Whatever the case, let us first congratulate you for your interest in this exciting subject! We hope these notes will help you on your way.

We encourage you to work through the "review" Chap. 2 first to familiarize yourself with all the necessary tools and concepts in math and physics. If you are unfamiliar with any of these concepts, you can look up more information online or in certain advanced high school text books (e.g. on Precalculus or Calculus). The concepts in Chap. 2 will be assumed as familiar in the following chapters.

The *Discussion* questions in the boxes can help guide you through learning new concepts; try to think hard about the questions posed there on your own before reading on in the text proceeding the boxes (where often the answers are given). For some of the *Discussion* questions, you can also check your answers in Appendix B.

We recommend skipping all of the barred sections on a first pass through the notes; they can always be revisited later. In general, don't be intimidated by the math in the notes, especially in Chap. 4; you don't need to fully grasp all equations you encounter along the way to end up with a good conceptual and intuitive understanding of relativity. In fact, once you understand most of the conceptual points relating to parallel transport, geodesics, and curvature (introduced in the *Discussion* questions 4.D, 4.E, and 4.F), you probably already have enough understanding to read through and get the gist of the more "fun" parts in Sect. 4.6 (including black holes, time machines, wormholes, and more!). The exercises that you will encounter along the way are meant as a guide to help you understand the concepts better and practice using them in actual applications. Some exercises should be fun to figure out, too— for example, you can calculate the special and general relativistic effects that GPS satellites need to take into account (Exercise 4.40), calculate how much more your feet age compared to your head (Exercise 4.41), figure out if you can outrun a light ray (Exercise 3.73), and a lot more. If you are feeling particularly adventurous, try doing the exercises with up to three stars, but beware—these can get *very* challenging!

Chapter 2
Fundamentals of Math and Classical Mechanics

2.1 Fundamentals of Math

Here, we will review some important concepts and tools in mathematics that we will need throughout our journey of learning relativity. We will cover the important concepts of Cartesian coordinate systems, changing coordinate frames, some basisc calculus involving derivatives, vectors and fields, and curvilinear coordinates.

2.1.1 Cartesian Coordinates

Coordinates are a way of labelling points in space. For example, standard *Cartesian coordinates* (x, y) label points on a plane (i.e. in two spatial dimensions), where x and y denote how far from the origin along the $x-$ or $y-$axis you need to travel to get to the point; see Fig. 2.1).

In three spatial dimensions, we can also introduce Cartesian coordinates (x, y, z). In general, we can also use a "numbering" notation for coordinates, i.e. (x^1, x^2) instead of (x, y) and (x^1, x^2, x^3) instead of (x, y, z). These numbered coordinates have the advantage that we can use the notation x^i to talk about them collectively, where i is then considered to be allowed to take on the values 1, 2 or 1, 2, 3, depending on the dimension of space.

2.1.2 Change of Coordinates

The Cartesian coordinate system (x, y) described above is not unique. For example, we can define a new (Cartesian) coordinate system (x', y'), related to the old coordinates as (see Fig. 2.2):

© Springer Nature Switzerland AG 2019
D. R. Mayerson et al., *Relativity: A Journey Through Warped Space and Time*,
https://doi.org/10.1007/978-3-030-18914-3_2

Fig. 2.1 Standard Cartesian coordinates (x, y) on the plane with point $(x, y) = (3, 4)$ indicated

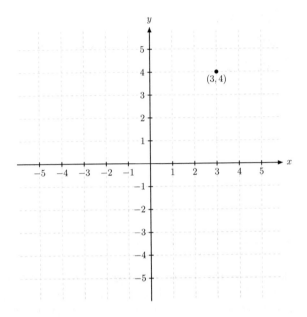

Fig. 2.2 Cartesian coordinate systems (x, y) and (x', y'), with $(x' = x + a, y' = y)$

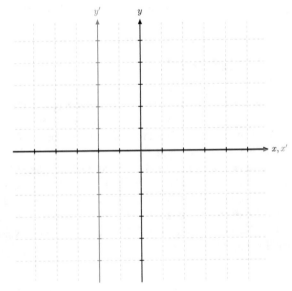

$$\begin{cases} x' = x + a \\ y' = y. \end{cases} \qquad (2.1)$$

This simply means that we have shifted the origin to the *left* by a. We could also define a new coordinate system by rotating our coordinate system by an angle θ[1] (see Fig. 2.3):

$$\begin{cases} x' = x \cos\theta + y \sin\theta, \\ y' = -x \sin\theta + y \cos\theta. \end{cases} \qquad (2.2)$$

We can also use matrix notation to write:

$$\begin{pmatrix} x' \\ y' \end{pmatrix} = \begin{pmatrix} x'^1 \\ x'^2 \end{pmatrix} = \begin{pmatrix} \cos\theta & \sin\theta \\ -\sin\theta & \cos\theta \end{pmatrix} \cdot \begin{pmatrix} x^1 \\ x^2 \end{pmatrix} = \begin{pmatrix} \cos\theta & \sin\theta \\ -\sin\theta & \cos\theta \end{pmatrix} \cdot \begin{pmatrix} x \\ y \end{pmatrix}. \qquad (2.3)$$

When we change coordinate systems from (x, y) to (x', y'), we are doing a *coordinate transformation*:

$$(x, y) \rightarrow (x', y'), \qquad x^i \rightarrow x'^i. \qquad (2.4)$$

In general, a *linear* coordinate transformation can be expressed in matrix notation:

$$\vec{x}' = M\vec{x} + \vec{a}, \qquad x'^i = \sum_j M^i{}_j x^j + a^i, \qquad (2.5)$$

where $\vec{x} = (x, y)$ is a vector, M is a matrix; the right equation is the same as the left one but written in components. More general coordinate transformations simply take the form:

$$\vec{x}' = \vec{x}'(\vec{x}), \qquad x'^i = x'^i(x^j), \qquad (2.6)$$

where we again give the same equation in vector and vector-component forms.

Exercise 2.1 Find the coordinate transformation that corresponds to first shifting the origin along the x-axis by a and then rotating by angle θ. Find the coordinate transformation that does the same but in opposite order. Is the end result the same in both cases?

---

[1]Note that we will always work with radians for angles. Remember that there are 360° in a full circle or 2π radians, so that the conversion between degrees and angles is $\frac{180}{\pi}\theta° \leftrightarrow \theta$ rad.

Fig. 2.3 Cartesian coordinate system (x, y) and rotated system (x', y') (given by (2.2)), with $\theta = 30° = \pi/6\,\text{rad}$

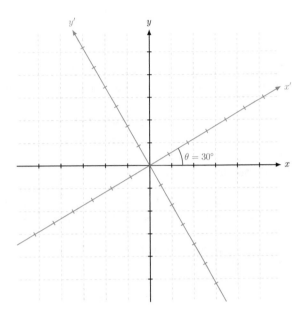

2.1.3　Calculus: Derivatives, Taylor Series

Formally, a derivative is defined through the limit:

$$f'(x) = \frac{d}{dx} f(x) = \lim_{\epsilon \to 0} \frac{f(x + \epsilon) - f(x)}{\epsilon}. \tag{2.7}$$

Graphically, the derivative of the function is the slope of the tangent line to the graph of $f(x)$ (see Fig. 2.4). This means the equation that gives the line tangent to the graph of $f(x)$ at a point $(x_0, f(x_0))$ is:

$$y = f'(x_0)(x - x_0) + f(x_0). \tag{2.8}$$

We will often need the following derivatives:

Function $f(x)$	Derivative $f'(x)$
c	0
x^n	nx^{n-1}
$\sin x$	$\cos x$
$\cos x$	$-\sin x$
$\tan x$	$\sec^2 x$
$\cot x$	$-\csc^2 x$
e^x	e^x
$\ln x$	$\frac{1}{x}$

$$\tag{2.9}$$

Fig. 2.4 The graph $y = f(x)$ of a specified function. The line tangent to the graph always has slope $f'(x)$

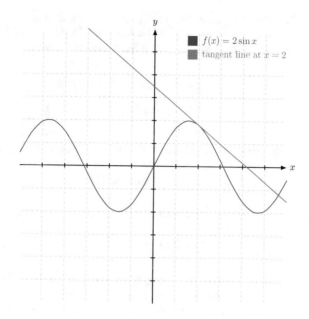

Also remember the all-important chain rule:

$$\frac{d}{dx} f(g(x)) = f'(g(x))g'(x). \tag{2.10}$$

Exercise 2.2 Calculate the derivative $f'(x)$ of:
(a) $f(x) = x^2$
(b) $f(x) = x^4 + 3x^3 + 2x + 5$
(c) $f(x) = \ln(\sin x)$
(d) $f(x) = \frac{1}{x}$
(e) $f(x) = g(x)/h(x)$

Intuitively, the derivative of a function at a given point x_0 tells you "how fast" the function is increasing (or decreasing) at that point. In principle, it tells you nothing about how the function will behave when you move away from the point x_0. However, for sufficiently smooth functions, if you know all of the (infinite) derivatives of the function at a given point, you know everything there is to know about that function. This is quantified in the *Taylor expansion* or *Taylor series* of a function:

$$f(x) = f(x_0) + (x - x_0)f'(x_0) + \frac{(x - x_0)^2}{2} f''(x_0) + \cdots + \frac{(x - x_0)^n}{n!} f^{(n)}(x_0) + \cdots . \tag{2.11}$$

For example, consider $f(x) = e^x$. We know that $f(x) = f'(x) = f''(x) = f^{(n)}(x) = e^x$, so if we want to perform the Taylor series around $x_0 = 0$, all we need to know is $f(0) = e^0 = 1$. Filling this in gives us:

$$f(x) = e^x = f(0) + xf'(0) + \frac{x^2}{2}f''(0) + \cdots + \frac{x^n}{n!}f^{(n)}(0) + \cdots$$

$$= 1 + x + \frac{x^2}{2} + \cdots + \frac{x^n}{n!} + \cdots = \sum_{n=0}^{\infty} \frac{x^n}{n!}. \tag{2.12}$$

In this case, we can calculate the entire (infinite) Taylor series of the function without much trouble. Often in physics, we will consider the expansion of a function around a point only to first or second order. This means:

$$f(x + \epsilon) \approx f(x) + \epsilon f'(x) + O(\epsilon^2), \tag{2.13}$$

$$f(x + \epsilon) \approx f(x) + \epsilon f'(x) + \frac{\epsilon^2}{2} f''(x) + O(\epsilon^3). \tag{2.14}$$

The $\approx$ sign tells us this is only an approximate equality. $O(\epsilon^n)$ means we are neglecting terms that are proportional to ϵ^n or higher powers of ϵ; if ϵ is small, these terms should be much smaller than the terms we are taking into account! For $f(x) = e^x$, we can approximate:

$$f(0 + \epsilon) = e^\epsilon \approx 1 + \epsilon + \frac{\epsilon^2}{2}. \tag{2.15}$$

Exercise 2.3 Test out how good the approximation in (2.15) is, by comparing the exact answer (e^ϵ) with the quadratic approximation, for $\epsilon = 0.001; 0.01; 0.1; 1$. What do you conclude?

Exercise 2.4 Use the formula (2.11) to prove that the Taylor series (around $x_0 = 0$) stops after a finite number of terms for the polynomial $f(x) = x^3 + 2x^2 + x + 2$, and the Taylor series just gives us back the function itself. Is this always true for a polynomial?

Exercise 2.5 Calculate the Taylor expansion of $\sin x$ and $\cos x$; your answers should be expressed in terms of an infinite sum. Now take the derivative (with respect to x) of both of these (infinite) sums, and rewrite the result in such a way that you see that $\frac{d}{dx}(\sin x) = \cos x$ and $\frac{d}{dx}(\cos x) = -\sin x$.

Exercise 2.6 Consider $f(x) = \frac{1}{1-x}$. Calculate the Taylor series for $f(x)$ around $x = 0$ to second order (i.e. up to x^2 terms). Compare the exact value of $f(x)$ to the value of the Taylor series to second order for the points $x = 0, 0.01, 0.1, 0.5, 1, 2$.

When we have a function of multiple variables, a *partial derivative* is simply a derivative with respect to one variable, treating the others as constants:

$$\partial_x f(x, y) = \frac{\partial}{\partial x} f(x, y) = f^{(1,0)}(x, y). \tag{2.16}$$

We will often use notation such as:

$$\partial_{x^i} x'(x^1, x^2) = \frac{\partial x'}{\partial x^i}(x^k). \tag{2.17}$$

For example, for $f(x, y) = x + xy$:

$$\partial_x f(x, y) = 1 + y, \qquad \partial_y f(x, y) = x. \tag{2.18}$$

Exercise 2.7 Calculate the partial derivative $\partial_x f(x, y)$ for the following functions $f(x, y)$:

(a) $f(x, y) = x^2$
(b) $f(x, y) = y^2$
(c) $f(x, y) = x^2 y^2$
(d) $f(x, y) = e^{x+y} + e^{xy}$

Exercise 2.8 Calculate, for $f(x, y) = e^x + \sin y + \sin(xy)$

(a) $\partial_x^2 f(x, y)$
(b) $\partial_x \partial_y f(x, y)$
(c) $\partial_y \partial_x f(x, y)$ (why is this the same as above?)
(d) $\partial_y^2 f(x, y)$

A function of multiple variables can also be Taylor expanded in multiple variables:

$$f(x, y) = f(x_0, y_0) + (x - x_0)\partial_x f(x_0, y_0) + (y - y_0)\partial_y f(x_0, y_0) + \cdots + \frac{(x - x_0)^n}{n!}\frac{(y - y_0)^m}{m!}\partial_x^n \partial_y^m f(x_0, y_0) + \cdots. \tag{2.19}$$

A simple example is $f(x, y) = x + y^2$. Here, $\partial_x f = 1$ while $\partial_y f = 2y$ and $\partial_y^2 f = 2$. (Are these the only derivatives we care about?) If we expand around the point $(x_0, y_0) = (1, 2)$, noting that e.g. $f(1, 2) = 5$, then we get:

$$f(x, y) = f(1, 2) + (x - 1)\partial_x f(1, 2) + (y - 2)\partial_y f(1, 2) + \frac{(y - 2)^2}{2}\partial_y^2 f(1, 2) \tag{2.20}$$

$$= 5 + (x - 1)(1) + (y - 2)(4) + \frac{(y - 2)^2}{2}(2) \tag{2.21}$$

$$= 5 + (x - 1) + (4y - 8) + (y^2 + 4 - 4y) \tag{2.22}$$

$$= x + y^2. \tag{2.23}$$

Exercise 2.9 Calculate the Taylor expansions of $f(x, y)$ around $(x_0, y_0) = (0, 0)$ for:

(a) $f(x, y) = e^x + e^y$
(b) $f(x, y) = e^{x+y}$
(c) $f(x, y) = e^x \sin y$

Using the operator $\vec{\nabla}$ defined as (see further on for vector notations):

$$\vec{\nabla} = (\partial_x, \partial_y), \qquad (2.24)$$

we can define the *gradient* of a function or scalar as the following vector:

$$\vec{\nabla} f(x, y) = (\partial_x f(x, y), \partial_y f(x, y)), \qquad \nabla_i f(x^j) = \partial_i f(x^j), \qquad (2.25)$$

and we can define the *divergence* of a vector as the following scalar:

$$\vec{\nabla} \cdot \vec{v} = \partial_x v^x + \partial_y v^y = \sum_i \nabla_i v^i = \sum_i \partial_i v^i. \qquad (2.26)$$

2.1.4 Vectors and Fields

In its most basic form, a vector can be thought of as an object that has a magnitude and direction. We can draw it as an arrow originating at the origin and pointing towards the point (x, y). Vectors can be written as having an arrow above them: $\vec{v}$. All vectors can be written as a linear combination of a *basis* of vectors. In the plane in Cartesian coordinates, $\{\vec{e}_x, \vec{e}_y\}$ is a basis of vectors, where e.g. $\vec{e}_x$ is the unit vector (i.e. having length one) pointing along the x axis. Every vector in the plane can then be written as a linear combination of these unit vectors:

$$\vec{v} = v^x \vec{e}_x + v^y \vec{e}_y = v^1 \vec{e}_1 + v^2 \vec{e}_2 = \sum_i v^i \vec{e}_i, \qquad (2.27)$$

where we have defined the quantities v^x, v^y (or v^i). In Fig. 2.5, $v^x = 3$ and $v^y = 4$. In Cartesian coordinates, we will take $\vec{e}_i$ to be unit vectors, but this is not necessarily true for other coordinate systems (see e.g. Exercise 2.20).

We can equivalently denote a vector by $\vec{v}$ or by its components (v^x, v^y). When we use the numbered notation for coordinates, we can also just write v^i, where again i can take on all as many values as we have coordinates. We will often use the notation v^i instead of (v^x, v^y) as it is more general and can also be used when we use other coordinate systems other than the (x, y) Cartesian coordinate system.

Fig. 2.5 A vector, originating at the origin and pointing towards $(3, 4)$. The unit vectors $\vec{e}_x, \vec{e}_y$ are also drawn

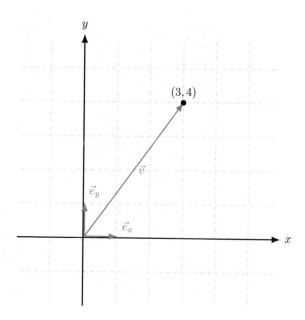

A *scalar field* is simply a number (a scalar) that we associate to every point in space—i.e.: a function.

A *vector field* is a vector that we associate to every point in space (see Fig. 2.6). Thus, its components v^i are functions of the coordinates.

We can take the dot product of any vector, defined as:

$$\vec{v} \cdot \vec{w} = v^x w^x + v^y w^y = v^1 w^1 + v^2 w^2 = \sum_i v^i w^i. \tag{2.28}$$

If we dot a vector with itself, this gives the magnitude of the vector squared:

$$\vec{v} \cdot \vec{v} = (v^x)^2 + (v^y)^2 = \sum_i v^i v^i = v^2. \tag{2.29}$$

We can generalize all of these notions easily to three spatial dimensions (by including a third coordinate $z = x^3$, a third vector component $v^z = v^3$, etc.)

2.1.5 Tensors

A scalar field is simply a number associated to each point in space. The value of the scalar field should only depend on the point in space and not on the coordinates used

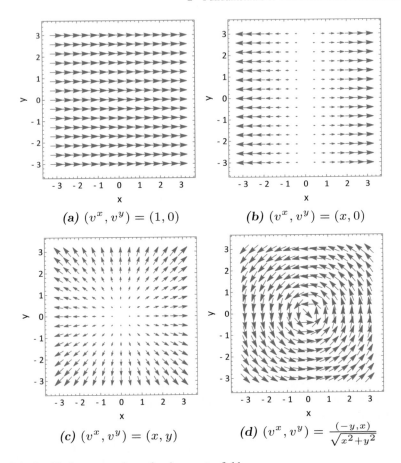

Fig. 2.6 Graphical representations of various vector fields

to describe the point. This means that, under a coordinate transformation $x \to x'$, a scalar field f should *transform* as:

$$f(\vec{x'}) = f(\vec{x}), \tag{2.30}$$

when we change coordinates from $\vec{x}$ to $\vec{x'}$.

A vector has direction and magnitude. These properties of a vector should not also not depend on what coordinates are used. This leads to the following transformation property of components of a vector:

$$v'^x = \frac{\partial x'}{\partial x} v^x + \frac{\partial x'}{\partial y} v^y, \qquad v'^y = \frac{\partial y'}{\partial x} v^x + \frac{\partial y'}{\partial y} v^y. \tag{2.31}$$

Exercise 2.10 Take a vector that points in the x-direction with some arbitrary magnitude, and work out the components when we take as coordinates (x', y'), using (2.31):

(a) The shifted coordinate system $(x', y') = (x + a, y)$.
(b) The shifted coordinate system $(x', y') = (x, y + b)$.
(c) The rotated coordinate system given by (2.2).

Convince yourself that the answers you found make sense and thus that (2.2) is correct.

We can also use the numbered coordinates to express the above transformation property:

$$v'^1 = \frac{\partial x'^1}{\partial x^1} v^1 + \frac{\partial x'^1}{\partial x^2} v^2, \qquad v'^2 = \frac{\partial x'^2}{\partial x^1} v^1 + \frac{\partial x'^2}{\partial x^2} v^2. \tag{2.32}$$

This can be written in compact form as:

$$v'^i = \sum_j \frac{\partial x'^i}{\partial x^j} v^j. \tag{2.33}$$

This formula has the advantage that it immediately generalizes to higher dimensions. We recognize that this is really just a matrix equation:

$$\vec{v}' = M \cdot \vec{v}, \qquad M = \begin{pmatrix} \frac{\partial x'}{\partial x} & \frac{\partial x'}{\partial y} \\ \frac{\partial y'}{\partial x} & \frac{\partial y'}{\partial y} \end{pmatrix}, \tag{2.34}$$

or, using numbered coordinate notation:

$$v'^i = \sum_j M^i{}_j v^j, \qquad M^i{}_j = \frac{\partial x'^i}{\partial x^j}. \tag{2.35}$$

A vector *field* is defined at every point in space, so we must be careful that the coordinates of the point at which we define the vector transform as well:

$$v'^i(x'^k) = \sum_j \frac{\partial x'^i}{\partial x^j} v^j(x^k), \qquad v'^i(x'^k) = \sum_j M^i{}_j v^j(x^k), \tag{2.36}$$

where $\frac{\partial x'^i}{\partial x^j}$ denotes the partial derivative of the function $x'^i(x^1, x^2, \ldots)$ with respect to x^j; this is to be evaluated at the point x^k.

Exercise 2.11 Take the (4) vector fields from Fig. 2.6 and the (3) coordinate transformations from the previous exercise, and work out the shifted vector fields in each case (or at least until you feel confident with it).

We can also simply *define* a vector field to be something that transforms as (2.36). This definition can then easily be generalized: a *tensor* field with two indices is defined to be a quantity with components T^{ij} that transform in the following way under coordinate transformations:

$$T'^{ij}(x'^a) = \sum_{k,l} \frac{\partial x'^i}{\partial x^k} \frac{\partial x'^j}{\partial x^l} T^{kl}(x^a), \qquad T'^{ij}(x'^a) = \sum_{k,l} M^i{}_k T^{kl} (M^T)_l{}^j, \qquad (2.37)$$

where M^T is the transpose of the matrix M.

Exercise 2.12 Consider two scalar fields $f(x^i)$, $g(x^i)$. Is the quantity defined by $\tilde{v}^i = (f, g)$ a vector field? Why (not)?

Exercise 2.13 Define a tensor with n indices by generalizing the definition (2.37).

Exercise 2.14 If you have two vectors v^i and w^j, is the product $T^{ij} = v^i w^j$ a tensor?

Exercise 2.15 How many independent components does the tensor T^{ij} have, if we are working in 3D? What about in general dimension d? What if the tensor is symmetric (so $T^{ij} = T^{ji}$), or antisymmetic (so $T^{ij} = -T^{ji}$)? How many independent components does a generic tensor $T^{ij\cdots}$ have with n indices in d dimensions?

Up to now, we have always considered vectors or tensors with indices "upstairs"—these are called *contravariant* indices. We can also place these "downstairs", and call them *covariant* indices. An object with one covariant index is defined as an object that transforms with the *inverse* coordinate transformation, i.e. the inverse of the matrix M:

$$K'_i(x'^l) = \sum_j \frac{\partial x^j}{\partial x'^i} K_j(x^l), \qquad K'_i(x^l) = \sum_j (M^{-1})_i{}^j K_j(x^l) \qquad (2.38)$$

Thus, a covariant tensor transforms with the *inverse* coordinate transformation $x^i(x'^j)$ instead of $x'^i(x^j)$.

Exercise 2.16 Generalize this definition to define a tensor with n contravariant indices and m covariant indices.

Exercise 2.17** Consider $\nabla_i f = \partial_i f$; prove that it transforms as (2.38). Also prove that the divergence of a vector is indeed a scalar.

Exercise 2.18 Convince yourself that the matrix defined by:

$$\tilde{M}_i{}^j = \frac{\partial x^j}{\partial x'^i}, \tag{2.39}$$

is indeed the inverse of the matrix $M^i{}_j$, i.e. that $M \cdot \tilde{M} = \tilde{M} \cdot M = \mathbb{I}$ (where $\mathbb{I}$ is the identity matrix).

Clearly, the position of the index is very important when we write down a tensor or vector, as it tells us how the tensor transforms under coordinate transformations!

2.1.6 Curvilinear Coordinates

Up until now, we have always used Cartesian coordinates to label points in space. Another well-known coordinate system on the plane are *polar coordinates*, which are labelled by a radius r (distance from the origin) and angle (from x-axis) θ (in radians).

The coordinate transformation relating polar coordinates to Cartesian coordinates is (see Fig. 2.7):

$$\begin{cases} x = r \cos \theta, \\ y = r \sin \theta. \end{cases} \tag{2.40}$$

Exercise 2.19 Figure out the inverse coordinate transformation, i.e. find (r, θ) in function of (x, y). What is the point $(x, y) = (3, 5)$ in polar coordinates? What is the point $(r, \theta) = (3, \pi/3)$ in Cartesian coordinates?

Exercise 2.20 The basis vectors $\vec{e}_i$ have a covariant index; use that information to find $\vec{e}_r$ and $\vec{e}_\theta$. Are these unit vectors (if $\vec{e}_x, \vec{e}_y$ are unit vectors)?

The analogue of polar coordinates in 3D is *spherical coordinates* (r, θ, ϕ), which are related to Cartesian coordinates as (see Fig. 2.8):

$$\begin{cases} x = r \sin \theta \cos \phi, \\ y = r \sin \theta \sin \phi, \\ z = r \cos \theta. \end{cases} \tag{2.41}$$

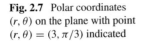

Fig. 2.7 Polar coordinates (r, θ) on the plane with point $(r, \theta) = (3, \pi/3)$ indicated

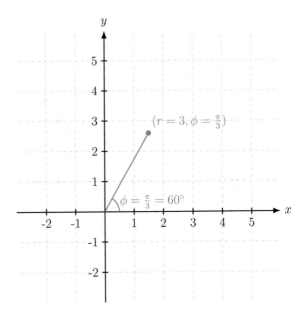

Exercise 2.21 The *range* for a coordinate is the set of values the coordinate is allowed to take. Clearly, for Cartesian coordinates, x, y, z can each take on any value between $-\infty$ and $+\infty$. What is the allowed coordinate range for polar coordinates (r, θ)? And spherical coordinates (r, θ, ϕ)?

Exercise 2.22 *Cylindrical* coordinates (ρ, θ, z) are another coordinate system in 3D; they are related to Cartesian coordinates as:

$$\begin{cases} x = \rho \cos \theta, \\ y = \rho \sin \theta, \\ z = z. \end{cases} \qquad (2.42)$$

Can you see why these are called *cylindrical* coordinates?

Exercise 2.23 Convince yourself that Cartesian (x, y) coordinates and polar coordinates do cover the *same* space, i.e. there are no points that you *can* describe in Cartesian coordinates but *cannot* describe with polar coordinates. Convince yourself the same about 3D Cartesian coordinates (x, y, z) and spherical coordinates (and cylindrical coordinates if you wish).

Fig. 2.8 A depiction of 3D spherical coordinates (r, θ, ϕ)

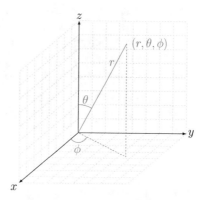

If we want to specify a certain subspace of or curve in e.g. the plane, we can write down equations involving the coordinates that specify what region of space we are interested in.

Exercise 2.24 Certain coordinate systems may be better adapted at describing different spaces. Draw the curve $x^2 + y^2 = 4$. What is this equation in polar coordinates?

Draw the line $x = 2y$. What is this equation in polar coordinates?

Can you figure out the equation(s) in polar coordinates for the subspace $x < 2$?

More Fundamentals of Math Exercises

Exercise 2.25 We have been very cavalier with (or more bluntly: ignored) various mathematical subtleties in the above discussion. A proper mathematical treatment would pay attention to these subtleties. However, for most applications in physics, such subtleties do not usually occur and we can safely ignore them. We will highlight some of these mathematical subtleties here with some example exercises.

(a) What is the value of θ in the origin ($x = y = 0$) for polar coordinates? The origin for polar coordinates is what is called a *coordinate singularity* and is an artifact of the coordinates used (i.e. there is nothing "special" going on at the origin; we could have chose the origin to be anywhere else instead).

(b) Do partial derivatives in different values always commute, i.e. $\partial_x \partial_y f = \partial_y \partial_x f$? Being very careful, calculate $\partial_x \partial_y f$ and $\partial_y \partial_x f$ for the function defined by $f(x, y) = xy(x^2 - y^2)/(x^2 + y^2)$ when $(x, y) \neq (0, 0)$ and $f(0, 0) = 0$.

(c) Calculate the Taylor series for $f(x) = 1/(1-x)$ at the point $x = 0$. Does the answer make sense for $x > 1$?

(d) Can you even begin to think of constructing a Taylor series for the function $f(x) = \exp(1/x)$ around $x = 0$? Why (not)? This function has what is called an *essential singularity* at $x = 0$.

Exercise 2.26**   Find the components of $\vec{\nabla}$ in polar coordinates. Your answer should be of the form $\vec{\nabla} = (f(r, \theta)\partial_r + g(r, \theta)\partial_\theta, h(r, \theta)\partial_r + l(r, \theta)\partial_\theta)$.

2.2   Classical Mechanics

In this section, we will review basic concepts in classical mechanics: position, velocity, acceleration; momentum, energy, force; Newton's laws; how physics in different reference frames are related. It is important to understand these concepts well in classical mechanics to be able to understand how the physics of (special) relativity contrasts with our everyday expectations of classical, Newtonian physics.

2.2.1   Position, Velocity, Acceleration

Let's say we have a particle that moves in the (x, y)-plane. This means its motion is completely determined by two functions $(x(t), y(t))$, that tells us the location of the particle at each instance in time t. We call:

$$\vec{x}(t) = (x(t), y(t)), \qquad x^i(t), \tag{2.43}$$

the *position* of the particle. The SI unit of position (more precisely: of each component of the position!) is meter (m).

If we know where the particle is at every moment, we can also find its *velocity vector*, which is simply the derivative with respect to time of the position:

$$\vec{v}(t) = \frac{d}{dt}\vec{x}(t) = \left(\frac{d}{dt}x(t), \frac{d}{dt}y(t)\right), \qquad v^i(t) = \frac{d}{dt}x^i(t). \tag{2.44}$$

The velocity of a particle is a vector: the velocity tells us *where* the particle is heading next (the direction of the vector), and *how fast* the particle is moving (magnitude of the vector). The SI units of velocity are meter/second (m/s).

Finally, two time-derivatives acting on the position gives us the *acceleration vector*:

$$\vec{a}(t) = \frac{d^2}{dt^2}\vec{x}(t) = \frac{d}{dt}\vec{v}(t) = \left(\frac{d^2}{dt^2}x(t), \frac{d^2}{dt^2}y(t)\right), \qquad a^i(t) = \frac{d^2}{dt^2}x^i(t). \quad (2.45)$$

The SI units for acceleration are meters/second-squared (m/s^2).

2.2.2 Momentum, Energy, Force

Let's say the particle that we considered above has mass m, with SI units kg. Then its *momentum* $\vec{p}$ is a vector, defined simply as mass times velocity:

$$\vec{p} = m\vec{v}, \qquad p^i = mv^i, \quad (2.46)$$

and its *kinetic energy* E_{kin} is given by the expression:

$$E_{kin} = \frac{1}{2}mv^2 = \frac{1}{2}m\vec{v} \cdot \vec{v} = \frac{1}{2}m\left((v^x)^2 + (v^y)^2\right). \quad (2.47)$$

Note that the momentum of a particle is a *vector* quantity (with magnitude and direction), while its energy is a *scalar* quantity (with no direction).

When we push or pull the particle, we exert a *force* on the particle. The force $\vec{F}$ is also a vector quantity with magnitude (how hard we push) and direction (in which direction we push). Forces are expressed in units of Newton (N), which can be expressed in fundamental units as $N = kg\,m/s^2$.

Some forces can be expressed as derivatives of a potential $V(\vec{x})$; these are called *conservative* forces:

$$\vec{F}_V = -\frac{d}{d\vec{x}}V(\vec{x}) = -\left(\frac{d}{dx}V(x, y), \frac{d}{dy}V(x, y)\right), \qquad (F_V)_i = -\nabla_i V(x^j).$$

$$(2.48)$$

When a particle is subjected to such a force, $V(\vec{x})$ is the *potential energy* of the particle:

$$E_{pot} = V. \quad (2.49)$$

The *total energy* of the particle is then:

$$E_{total} = E_{kin} + E_{pot}. \quad (2.50)$$

Exercise 2.27   Consider a car moving on a trajectory in the (x, y) plane. Choose and draw such a trajectory. Describe and/or calculate:

(a) Its position $x^i(t)$; velocity $v^i(t)$; acceleration $a^i(t)$. Pay special attention to the acceleration along curved parts of the trajectory.
(b) Its momentum $p^i(t)$ and its kinetic energy E_{kin}. Does the momentum necessarily change along curved parts of the trajectory? What about the kinetic energy?

2.2.3 Newton's Laws

Newton's first law, the law of *inertia*, tells us that a particle with no forces acting on it will have constant momentum. If it is at rest, it will remain at rest. If it is travelling with constant velocity, it will continue travelling with that same constant velocity.

Newton's second law tells us how a particle reacts to a force acting on it. It reacts by changing its velocity in this way:

$$\vec{F} = m\vec{a} = \frac{d}{dt}\vec{p}, \qquad F^i = ma^i, \tag{2.51}$$

in other words, the force applied to the particle is *precisely* the change over time in the particle's momentum.

Exercise 2.28   Use Newton's second law to prove Newton's first law, thus proving that the first law is actually redundant.

Newton's third law tells us that for each action (force), there is an equal and opposite reaction (force). In other words, if I push the wall with a force $\vec{F}$, the wall also pushes me with force $-\vec{F}$; see Fig. 2.9.

A consequence of Newton's laws is that *total momentum is always conserved*. If we have a system of n particles, who are allowed to interact with each other but are otherwise left alone, we have:

$$\sum_{k=1}^{n} \vec{p}_k = const, \qquad \sum_{k=1}^{n} (p_k)^i = const \tag{2.52}$$

Another consequence of Newton's laws is that *total energy is always conserved*. Again, for our system of n interacting particles, this means that:

$$\sum_{k=1}^{n} E_k = const. \tag{2.53}$$

Fig. 2.9 For every force, there is an equal and opposite force

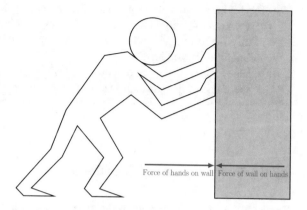

Force of hands on wall | Force of wall on hands

In fact, *conservation of momentum* and *conservation of energy* should be seen as "more fundamental" than Newton's laws; starting from conservation of momentum and energy, we can derive Newton's laws. When we consider relativity, we will continue to demand conservation of energy and momentum, but Newton's laws will strictly speaking not be valid anymore.

Exercise 2.29 Show that:

(a) Conservation of momentum and energy follow from Newton's laws.
(b) Newton's laws follow from conservation of momentum and energy.

Exercise 2.30 Two balls are travelling along the x-axis; one ball has mass M and velocity V, the other mass m and velocity v. They collide. After the collision, they now have velocities V' and v'.

(a) Use conservation of energy and momentum to find that:

$$v' = \frac{(m - M)v + 2MV}{m + M}, \qquad V' = \frac{(M - m)V + 2mv}{M + m}. \qquad (2.54)$$

(Also interpret the second solution you get in the process of solving the equations!)
(b) Take the limit where M is much bigger than m. What happens?
(c) Take the limit where $M = m$. What happens?

Exercise 2.31** Now consider two balls travelling in a plane; one ball with mass M and velocity $\vec{V}$, the other with mass m and velocity $\vec{v}$. They collide and again move after the collision with velocities $\vec{V'}$ and $\vec{v'}$.

(a) Without actually solving equations, can you convince yourself why con-
    servation of energy and momentum does not give enough information to
    fully determine $\vec{V}'$ and $\vec{v}'$? Hint: How many unknowns and how many
    equations do you have?
(b) What extra information do you need? Hint: Try to understand the case first
    where one of the particles is at rest before the collision, e.g. $\vec{v} = 0$.

2.2.4 Newton's Law of Gravitation

Newton's law of gravitation states that two masses M and M', which sit at a distance
R from each other, will exert a force on each other, of which the magnitude is given
by:

$$F_{grav} = -\frac{GMM'}{R^2}.$$ (2.55)

The direction of the force is on the line that connects the two masses, and the minus
sign indicates that the force is *attractive*, i.e. the masses experience a force pulling
them together. G is Newton's gravitational constant and is given by:

$$G = 6.67408 \cdot 10^{-11} \frac{m^3}{kg\ s^2}.$$ (2.56)

A consequence of Newton's law is that a small mass m on earth experiences an
approximately uniform gravitational force given by:

$$\vec{F}_{grav} = -m\,\vec{g},$$ (2.57)

where the magnitude of $\vec{g}$ at the surface of the Earth is given by:

$$g = |\vec{g}| = 9.8 \frac{m}{s^2},$$ (2.58)

and its direction is always pointing "down", i.e. towards the center of the Earth.

If we align $\vec{g}$ to point towards the negative y-axis, then the potential associated
with the force (2.57) is:

$$V_{grav} = mgy.$$ (2.59)

The total energy of a particle in the gravitational field of the Earth is then:

$$E_{total} = \frac{1}{2}m\vec{v}^2 + mgy.$$ (2.60)

Exercise 2.32 Given (2.55) and (2.57), derive a relation between the mass and radius of the Earth. Then calculate the radius and mass of the earth using the additional fact that $g = 9.77$ when we are $10\,km$ above the surface of the Earth.

Exercise 2.33 If we have a particle in the Earth's gravitational field, is this particle's energy conserved? How about its momentum? How does this make sense, given the previous claims of conservation of total energy and momentum? Also think about how Newton's laws can continue to make sense.

Exercise 2.34 You are standing on the Earth and throw a ball. When you throw it, the ball is at $y = 1m$. You throw the ball horizontally, so when you release it, it has velocity $v^x = 10m/s$ and $v^y = 0m/s$. Describe the ball's trajectory.

Exercise 2.35* Same situation as above, only now you throw the ball so that it has the same initial *magnitude* of its velocity, but its initial velocity *direction* makes an angle θ with the ground. Describe the ball's trajectory. At what angle should I release the ball to throw it the furthest?

2.2.5 Physics in Different Reference Frames

A fundamental principle and deep insight in physics is that *physics should be the same for every inertial observer*. This *principle of equivalence* may seem rather obvious, but it has very far-reaching consequences. In fact, we will see that different interpretations of this principle will lead us from classical mechanics inevitably to special and general relativity.

What do we mean precisely by the statement that "physics" should be the same for every inertial observer? It means that every inertial observer should see the same physical laws being obeyed. An "inertial observer" is an observer where no external forces are acting on. In Newtonian (classical) mechanics, all possible inertial observers are then necessarily moving at constant velocity with respect to each other.

Consider an inertial observer, Mr. A, who sits at the origin in a coordinate system (x, y). Another inertial observer, Mrs. B, sits at the origin in a coordinate system given by:

$$\begin{cases} x' = x - ut, \\ y' = y. \end{cases} \tag{2.61}$$

where u is some constant.

Exercise 2.36   From the perspective of Mr. A, what is Mrs. B's velocity? From Mrs. B's perspective, what is Mr. A's velocity?

Now, Mr. A observes a particle with mass m and velocity $\vec{v}$ fly by. Thus, according to Mr. A, the particle's momentum is $\vec{p} = m\vec{v}$.

Exercise 2.37   What is the momentum of the particle according to Mrs. B? Convince yourself that this is not in contradiction with conservation of momentum. Moreover, convince yourself that momentum is conserved in Mrs. B's reference frame *if and only if* momentum is conserved in Mr. A's reference frame.

We see that momentum itself is a quantity that is dependent of observer, but *conservation of momentum* is a principle that is independent of who is observing.

Exercise 2.38   Consider the particle's (kinetic) energy in Mr. A and Mrs. B's frames. What can you say about the principle of conservation of energy?

2.2.6 Vector Cross Products and Angular Momentum

We will not really need these concepts in the rest of these notes, but mentioning angular momentum completes our discussion of conservation laws in classical mechanics.

For two vectors in 3D, we can define a vector cross product as a vector given by:

$$\vec{v} \times \vec{w} = \left(v^y w^z - v^z w^y,\, v^z w^x - v^x w^z,\, v^x w^y - v^y w^x\right). \tag{2.62}$$

The vector $\vec{v} \times \vec{w}$ will have magnitude $|\vec{v}||\vec{w}| \sin \theta$ where θ is the angle between the vectors (see Fig. 2.10) and $\vec{v} \times \vec{w}$ will be in a direction perpendicular to both $\vec{v}$ and $\vec{w}$; which direction (as there are two possibilities that are perpendicular to $\vec{v}, \vec{w}$) can be obtained by the *right-hand rule*.

If we are considering two vectors $\vec{v}, \vec{w}$ in the plane, their cross product $\vec{v} \times \vec{w}$ will always be aligned with the z-axis in 3D; this is no longer a 2D vector (but can be thought of as a scalar from the 2D perspective). Note that a cross product of a vector with another vector parallel to it is zero, and in particular $\vec{v} \times \vec{v} = 0$.

Fig. 2.10 Two vectors $\vec{v}, \vec{w}$ and their cross product $\vec{v} \times \vec{w}$

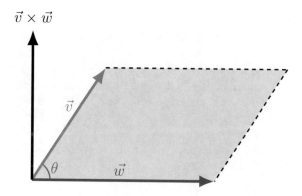

Exercise 2.39*   The ϵ tensor in 3D is a tensor with three indices: ϵ_{ijk}. This tensor is totally antisymmetric under interchanging a pair of indices, so e.g. $\epsilon_{ijk} = -\epsilon_{jik}$.

(a) We set $\epsilon_{123} = 1$. Does this completely determine all components of ϵ_{ijk}?
(b) Convince yourself that the numbered coordinate form of the cross product is given by:

$$(\vec{v} \times \vec{w})^i = \sum_{j,k} \epsilon_{ijk} v^j v^k. \tag{2.63}$$

If the position of a particle is the vector $\vec{r}$, the the angular momentum of a particle is the vector defined by:

$$\vec{L} = \vec{r} \times \vec{p}. \tag{2.64}$$

Newton's laws tell us that the total angular momentum of a system is conserved. Conservation of angular momentum is another principle that is very important in physics.

Exercise 2.40   Sit down on a swiveling chair. Swivel around and see how your velocity changes when you either tuck in or stretch out your legs. Can you understand this from the principle of conservation of angular momentum?

More Classical Mechanics Exercises

Exercise 2.41*** **The Lagrangian Formalism** It is often difficult to describe
the motion of a system with constraints. For example, consider the motion
of a particle confined to lie on top of a dome (half-sphere) of radius R in
the presence of gravity. It may be clear that the motion of the particle will
qualitatively be towards the ground, but can you describe the motion precisely
of the particle, released at an arbitrary point on the dome?

The *Lagrangian formalism* provides an elegant solution to this problem.
The Lagrangian is given by:

$$L(q^i) = T(q^i) - V(q^i), \tag{2.65}$$

where T is the kinetic energy (i.e. E_{kin}), $V(q)$ is the potential energy, and
q^i are called the *generalized coordinates*. These are the *free* parameters that
describe the particle; in other words, the generalized coordinates *already take
the constraints of the system into account*. In the example of the particle on the
dome, the generalized coordinates could be θ, ϕ of the dome (since the particle
is allowed to be anywhere on the dome). Note that the number of generalized
coordinates is always (number of freedoms or coordinates) minus (number of
constraints)—in our case, the particle has three coordinates and one constraint
(lying on the dome), so the number of generalized coordinates must be two.
For our particle, the kinetic energy in terms of $\theta(t), \phi(t)$ is:

$$T[\theta(t), \phi(t)] = \frac{1}{2}R^2(\dot{\theta}^2 + \sin^2\theta\dot{\phi}^2), \tag{2.66}$$

where $\dot{\ }$ denotes differentiation with respect to the time t. The potential energy
is (assuming the z-direction gives the height and $z = r\cos\theta$):

$$V[\theta(t), \phi(t)] = gR\cos\theta. \tag{2.67}$$

The Lagrangian is thus:

$$L[\theta(t), \phi(t)] = \frac{1}{2}R^2(\dot{\theta}^2 + \sin^2\theta(t)\dot{\phi}^2) - gR\cos\theta. \tag{2.68}$$

The Lagrangian formalism now tells us that the motion of the system is entirely
determined by the equation(s):

$$\frac{d}{dt}\left(\frac{\partial}{\partial\dot{q}^i}L[q^j, \dot{q}^j]\right) - \frac{\partial}{\partial q^i}L[q^j, \dot{q}^j] = 0, \tag{2.69}$$

which is a separate equation for each i. For our system above, these equations are:

$$R^2 \ddot{\theta} - R^2 \sin\theta \cos\theta \dot{\phi}^2 - gR \sin\theta = 0, \tag{2.70}$$

$$R^2 \sin^2\theta \, \ddot{\phi} + 2R^2 \sin\theta \cos\theta \, \dot{\theta} \, \dot{\phi} = 0. \tag{2.71}$$

While still formidable equations to solve, these two equations do *completely* determine the motion of our dome-constrained particle.

(a) Why have we ignored the mass of the particle m? Will it contribute any-where in the relevant equations determining the motion of the particle?

(b) Check that (2.66) and (2.67) are indeed the correct expressions for T and V in terms of $\theta(t)$, $\phi(t)$ for our constrained particle.

(c) Check that (2.70)–(2.71) are the right equations that follow from (2.69).

(d) Use (2.71) to find that:

$$R^2 \sin^2\theta \, \dot{\phi} = L, \tag{2.72}$$

where L is a constant. What does L represent physically?

(e) Now use L to eliminate derivatives of ϕ, and obtain the equation:

$$R^2 \ddot{\theta} = \frac{L^2}{R^2} \cot\theta \csc^2\theta + gR \sin\theta. \tag{2.73}$$

(f) This last equation is still quite hard to solve. Let's assume $\theta(t) = \pi/2 - \tilde{\theta}(t)$, where $\tilde{\theta}(t)$ is small. (What does this mean physically for the particle's location?) Explain that the equation above is approximated by:

$$R^2 \ddot{\tilde{\theta}} = -gR - \frac{L^2}{R^2} \tilde{\theta} + \cdots . \tag{2.74}$$

(g) Solve this differential equation for $L = 0$ and $L \neq 0$ separately. (Why do we need to separate these cases?) Identify the physical meaning of all of the integration constants you get. Finally, check that the approximation that $\tilde{\theta}(t)$ is small makes sense (and describe in which cases it doesn't).

Exercise 2.42*** More Lagrangians Use the Lagrangian formalism to find the motion of the following systems:

(a) (Atwood's machine.) Consider two masses m_1 and m_2, connected by a rope that hangs over a (frictionless) disc. The rope is sufficiently strong that its length remains constant. Use the Lagrangian formalism to obtain the equation of motion for the system. (Hint: There is only one generalized coordinate!) Could you have obtained the same result by other, "elementary" means?

(b) Two equal masses m are at opposite ends of a rigid (weightless) rod of length l. The center of this rod is at a fixed point. Find appropriate generalized coordinates and express the kinetic and potential energy for this system.

(c) Two equal masses m are at opposite ends of a rigid (weightless) rod of length l. The center of this rod is fixed on a given circle of radius a. Find appropriate generalized coordinates and find the kinetic and potential energy for this system.

(d) Find the Lagrangian for a spherical pendulum, i.e. a mass point suspended by a rigid weightless rod.

Chapter 3
Special Relativity

3.1 The Speed of Light

In the beginning of the twentieth century the general feeling was that physics was near "complete": we understood how things moved according to (classical) mechanics, which was codified in mathematical laws by Newton. This even included an understanding of gravity in the form of Newton's gravitational law. These simple and elegant mathematical laws explained essentially all of the known physical phenomena: from the astrophysical motion of the stars and planets (e.g. Kepler's laws) to the everyday mechanics experiments one could perform here on Earth.

In hindsight, we know the laws of classical mechanics are not really the "correct" laws of physics. We need relativity (special and general) to correctly describe our physical reality. How, then, could physicists at the beginning of the twentieth century think they were "almost done" with physics?

The reason is quite simple, and requires us to remember *who* is doing the experiments and the inference of physical laws. We, the experimenters, come burdened with what is perhaps the most insidious of obstacles standing in the way of scientific progress: common sense. Common sense tells us that if we throw a rock up, it will come down. Common sense tells us the laws of classical mechanics *make sense*: they describe the world precisely as we experience it, time and time again. But common sense and our everyday experiences don't prepare us for the "extremes" of physics! We never actually travel even remotely close to the speed of light. There is no fundamental reason to expect that our intuition, which is based on everyday experiences with relative velocities, can be extrapolated to velocities so large as (near) the speed of light.

Incidentally, we *are* quite sure that our intuition does not fail for everyday experiences. Thus, if we replace classical mechanics with another theory that will correctly describe high velocities (like special relativity), then that theory had better give the same results as classical mechanics when we consider low velocities! Checking that a "new" theory gives "old" results in this fashion is always the first consistency

© Springer Nature Switzerland AG 2019
D. R. Mayerson et al., *Relativity: A Journey Through Warped Space and Time*,
https://doi.org/10.1007/978-3-030-18914-3_3

check to do on the theory. We will make sure that special relativity gives the same predictions for everyday, low-speed physics.

The failure of pre-20th century classical mechanics started to become apparent when new methods and technology allowed experimenters to perform experiments that probed these "extremes" of physics. For special relativity, a key experiment was the Michelson-Morley experiment: this probed a fundamental underlying assumption of Newtonian physics and demonstrated that it completely and utterly failed to describe reality—leading necessarily to an entirely new description and understanding of physics and the spacetime in which it lives.

3.1.1 Electrodynamics and the Necessity of the Aether

The basic laws of Newtonian mechanics were known and well understood since Newton (obviously) in the 17th century. The same can not really be said about electromagnetism. Magnets were known and used in e.g. compasses, while electricity started to be used seriously in the 18th century; and while certain experiments showed a link between electricity and magnetism, true understanding of electricity and magnetism came by laborious experiments and insights in the 19th century, culminating in Maxwell's understanding of the two phenomena as intricately linked together in a theory of *electromagnetism*. It is important to realize that the laws of electromagnetism were a result of many years of countless experiments by different physicists—this is also why different laws of electromagnetism have different names, like *Ampère's law*, or *Gauss' law*. Maxwell had the insight in 1864 to bring these separate laws and phenomena together in a clear and unified theory of electromagnetism. Maxwell also noted that the equations of electromagnetism predict the existence of *electromagnetic waves*, which must travel at precisely the speed of light—so he proposed that light was, indeed, precisely an electromagnetic wave!

Newtonian physics and Maxwell's electromagnetism clashed in a violent way. As we saw, Newtonian physics hinged on the *principle of equivalence*—that different inertial observers were equivalent and should see the same laws of physics; translating between the reference frames of different inertial observers should be done with simple "Galilean" coordinate transformations given by (2.6.1). However, Maxwell's equations of electromechanics *did not transform correctly*, i.e. if one inertial observer saw electromagnetism, the transformation (2.6.1) together with the laws of electromagnetism necessarily implied that another inertial observer saw *different* laws. In other words, electromagnetism violated the principle of relativity.

To solve this violent clash, Maxwell and others took inspiration from another phenomenon associated with light: light travels at different speeds in different mediums—for example, light travels slower in water than in air. The existence of an invisible and barely detectable *aether* was postulated to exist everywhere in (what was thought to be) free space; this aether would be the medium through which light travels when it was not travelling through other (known) matter such as air or water. Conceptually, the aether provided a *preferred reference frame* for light and electromechanics: Maxwell's laws would be valid in the rest-frame of the aether; but now it made sense that in rest frames that were moving relative to the aether, the laws of electromechanics would seem different.

However, if there really is such a thing as the aether, it should have noticeable consequences, which can be tested by experiment! One such thing is the "aether frame-dragging": the earth moves around the sun and so does not always move at constant velocity. Thus, if there is an aether permeating all of space, the earth should be moving through the aether at different velocities with respect to the aether's rest frame at different moments. Electromagnetism, and therefore the speed of light, should then be different at different times, and at different directions. This different in the speed of light due to the aether frame-dragging is precisely what the famous Michelson-Morley experiment set out to measure.

3.1.2 The Michelson-Morley Experiment

We will describe a simplified version of the Michelson-Morley experiment. First, we will consider an apparatus, that shoots out light at $t = 0$ from the origin in two directions at once: up (towards positive y, staying at $x = 0$) and right (towards positive x, staying at $y = 0$). The light then reflects on mirrors situated at $x = L$ and $y = L$ and returns to the apparatus at the origin. See Fig. 3.1.

Exercise 3.1 If light travels at a speed c shoots in both directions at $t = 0$, at what time do the "vertical" and "horizontal" light beams return to the apparatus? Is this the same time for both directions?

The experiment we just did had the speed of light being c in every direction; this means we were at rest with respect to the aether. Let's see what happens when

Fig. 3.1 The Michelson-Morley experiment at rest

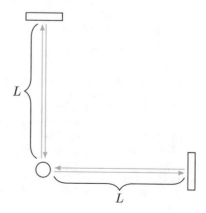

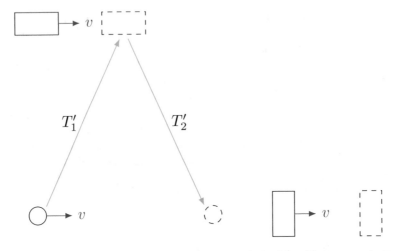

Fig. 3.2 The Michelson-Morley experiment, moving towards the right with constant velocity v

the apparatus is moving with respect to the aether. Let's consider the entire apparatus moving at constant velocity v in the positive x direction; this means the emitting/measuring apparatus (at the origin at $t = 0$) as well as the mirrors on which the light reflects—see Fig. 3.2.

Exercise 3.2   First, we consider the light travelling in the x-direction.

(a) If the speed of light was c when the apparatus was at rest, what is the speed of the light emitted in the x-direction when the apparatus is moving to the right?

(b) Find T_1, the time elapsed between the emitting of the light and the reflection of the light on the right mirror. Then find T_2, the time elapsed between the reflection of the light and the detection of the light back at the apparatus.

(c) We have now found $T_{x,total} = T_1 + T_2$. Use a Taylor expansion to prove that

$$T_{x,total} = \frac{2L}{c}\left(1 + \frac{v^2}{c^2}\right),\qquad(3.1)$$

when v is much less than c (i.e. $v \ll c$).

Exercise 3.3   Now, we consider the light travelling in the y-direction.

(a) Find T_1', the time elapsed between the emitting of the light and the reflection of the light on the upper mirror. Hint: the answer is *not* L/c!

(b) Find $T_{y,total}$ and prove that

$$T_{y,total} = \frac{2L}{c}\left(1 + \frac{v^2}{2c^2}\right).\qquad(3.2)$$

Exercise 3.4 We have thus found that the difference in time travelled is

$$\Delta T_{total} = T_{x,total} - T_{y,total} = \frac{L}{c} \frac{v^2}{c^2}. \tag{3.3}$$

Use $c = 3 \cdot 10^8$ m/s to find a numerical estimate for ΔT_{total} for realistic ("everyday") values of L, v.

The conclusion is clear: if the speed of light is c only in the rest-frame of the aether, then when we change reference frames, we should be able to measure a non-zero ΔT_{total} due to the differences in speed of light in the x- and y-direction. The earth travels very quickly around the sun, so we expect that at different times and different places on earth, we would have different relative velocities with respect to the aether. In other words, depending on when and where we perform this experiment, we should find different, non-zero values for ΔT_{total}.

What did the experiment actually find? $\Delta T_{total} = 0$. Always. The speed of light is *always* the same. The conclusion: there is no aether! Michelson received the Nobel prize in 1907 for his interferometer experiments.

3.2 Basic Relativistic Physics

The aether was postulated as a fix to the apparent violent clash between electromechanics and Newtonian physics. Experiments such as the Michelson-Morley experiment in 1887 attempted and failed to measure effects on the speed of light due to the aether; the speed of light was *always* the same! It was clear that the aether "fix" did not work. The stage was set for Einstein in 1905—his *annus mirabilis*, when he published four ground-breaking, tremendously important papers—to introduce his theory of special relativity. Einstein's original paper introduces special relativity and the Lorentz transformations as a way to save the principle of equivalence by taking seriously a prediction from electrodynamics: the speed of light is always constant, for any observer. The theory of special relativity, as he argued, completely explained how electrodynamics and the principle of equivalence could exist together. This made the existence of the aether completely superfluous.

3.2.1 Principles of Relativity

Einstein formulated the theory of special relativity as the mathematical interpretation of two fundamental physical principles.

The first principle is *the principle of relativity*: all inertial observers are equivalent and should see the same laws of physics.

Exercise 3.5 Is this the same as Newtonian physics? How about Newtonian physics with the aether?

The second principle is that *the speed of light is a universal constant*. It does not matter what inertial frame you are in—you will measure the same value for the speed of light.

3.2.2 Relativistic Reference Frames and Lorentz Transformations

We have two observers moving at a relative speed u with respect to each other. Our current, Newtonian way of understanding the difference in reference frames is the coordinate transformation (this is (2.61)):

$$\begin{cases} t' = t, \\ x' = x - ut, \\ y' = y, \\ z' = z. \end{cases} \tag{3.4}$$

Note that we included the transformation of the time t (as well as the third spatial dimension z). No matter what coordinate transformation we do, the time coordinate does not change. This is a key feature of Newtonian physics: there is a notion of *absolute time* that all observers can agree on. Things that happen at the same point in time for one observer will also always happen at the same time for any other observer.

However, an immediate consequence of the two principles of relativity above is that (3.4) must be wrong! If one observer measures the speed of light to be c in his reference frame, (3.4) immediately tells us that the other observer will measure the speed of light to be $c - u$.

The correct way to perform a coordinate transformation between two observers moving at a relative speed u is called the *Lorentz transformation*, and is given by:

$$\begin{cases} t' = \gamma\left(t - \dfrac{u}{c^2}x\right), \\ x' = \gamma\,(x - ut), \\ y' = y, \\ z' = z. \end{cases} \quad\Leftrightarrow\quad \begin{cases} ct' = \gamma\left(ct - \dfrac{u}{c}x\right), \\ x' = \gamma\left(x - \dfrac{u}{c}ct\right), \\ y' = y, \\ z' = z. \end{cases} \qquad \gamma = \sqrt{\dfrac{1}{1 - \dfrac{u^2}{c^2}}}. \tag{3.5}$$

In a sense, this is the most fundamental equation in special relativity, and everything else in the theory is simply exploring the consequences of this. In matrix notation:

$$\begin{pmatrix} ct' \\ x' \\ y' \\ z' \end{pmatrix} = \begin{pmatrix} \gamma & -\gamma\frac{u}{c} & 0 & 0 \\ -\gamma\frac{u}{c} & \gamma & 0 & 0 \\ 0 & 0 & 1 & 0 \\ 0 & 0 & 0 & 1 \end{pmatrix} \cdot \begin{pmatrix} ct \\ x \\ y \\ z \end{pmatrix}. \tag{3.6}$$

Exercise 3.6 Try to understand γ a bit better. What values can it take? When does it take the extreme values? Try to see why (3.5) does not make sense for velocities larger than the speed of light.

Exercise 3.7 A first sanity check of (3.5) is to understand the Newtonian limit: as we mentioned before, special relativity should match the predictions of classical mechanics at low velocities. Concretely, this means (3.5) should match (3.4) when u is much smaller then c. A useful Taylor expansion can be:

$$\sqrt{\frac{1}{1-\epsilon^2}} \approx 1 + \frac{\epsilon^2}{2}, \tag{3.7}$$

which is valid when ϵ is small.

(a) Use a Taylor expansion (what is the "small" variable in the expansion?) on the transformation (3.5) to prove that it reduces to (3.4) when $u \ll c$. *Hint*: it is useful to work with the combination ct (and ct') instead of just t.
(b) Using the next order in the Taylor expansion, find what the first order correction is to (3.4). For everyday velocities u and $c = 3 \cdot 10^8$ m/s, estimate this correction.

Exercise 3.8 Let's investigate how velocities change. Consider a constant velocity v in the x-direction (in the (t, x, y) frame). If v is constant, it can be measured by simply calculating $v = \Delta x / \Delta t$.

(a) Find v', the velocity in the (t', x', y') frame. Prove that:

$$v' = \frac{v - u}{1 - \frac{uv}{c^2}}. \tag{3.8}$$

(b) If u and v are both much smaller than c, convince yourself that this expression is consistent with Newtonian physics.
(c) If $v = c$, what is v'? Is this what you expect?
(d) What happens with v' when $u = c$?

Exercise 3.9** Let's consider a successive chain of reference frames, each moving with velocity u with respect to the previous frame. In the initial frame, we have a velocity v_0. In the first frame in the chain, we have the transformed velocity v_1 given by (3.8). In frame n, we have the transformed velocity v_n. Define $\tilde{v}_n = v_n/c$ and $\tilde{u} = u/c$. We have:

$$\tilde{v}_n = \frac{(1 + \tilde{v}_0)(1 - \tilde{u})^n - (1 - \tilde{v}_0)(1 + \tilde{u})^n}{(1 + \tilde{v}_0)(1 - \tilde{u})^n + (1 - \tilde{v}_0)(1 + \tilde{u})^n}, \tag{3.9}$$

(a) Check by hand that (3.9) is correct for $n = 1, 2, 3$.
(b) Prove that (3.9) is true for all n. Use induction, i.e. prove that if (3.9) holds for n, then it holds for $n + 1$. Compare (3.9) to the expression Newtonian physics would have predicted for v_n.
(c) Let's put $v_0 = 0$. Define β by $\tanh \beta = \tilde{u}$ and analogously $\tanh \beta_n = \tilde{v}_n$. Use the formula relating arctanh to ln's and the formula expressing cosh, sinh in terms of exponentials to prove that:

$$\beta_n = -n\beta. \tag{3.10}$$

The quantity β is called the *rapidity*; velocities are no longer additive in special relativity but rapidities are.

3.2.3 Spacetime Diagrams

A convenient graphical tool to depict things in special relativity are *spacetime diagrams*. Only considering t and x coordinates, we can draw x on the x axis and ct on the y-axis. In such a diagram, light always travels at a 45° angle, i.e. along the line $y = x$ or $x = ct$ (see Fig. 3.3).

We can also depict multiple relativistic reference frames in one spacetime diagram. To do this, let us figure out how the t'- and x'-axes should look like. From the Lorentz transformation (3.5), we see that the x'-axis, with equation $ct' = 0$, is at:

$$ct' = 0 \leftrightarrow x = (u/c)^{-1}(ct). \tag{3.11}$$

On the other hand, the ct'-axis, with equation $x' = 0$, is at:

$$x' = 0 \leftrightarrow x = u/c(ct). \tag{3.12}$$

Fig. 3.3 Simple spacetime diagram with light originating at $t = x = 0$

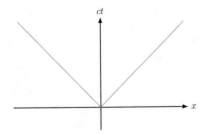

Fig. 3.4 A spacetime diagram with two reference frames (ct, x) and (ct', x'). Note that $\tan \theta = u/c$, where u is the relative velocity between the two reference frames

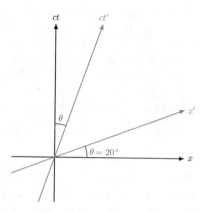

You can convince yourself that the x'-axis is at an angle α with the x-axis and the ct'-axis is at an angle $-\alpha$ with the ct-axis with:

$$\tan \alpha = \frac{u}{c}, \qquad (3.13)$$

see Fig. 3.4. Note that the factor γ means the ct' and x' axes are distorted with respect to the (ct, x) coordinate system; this makes calculating time and distance intervals in the primed system slightly more complicated.

We can use spacetime diagrams to graphically represent and clarify many things in special relativity, as we will see.

3.2.4 Length Contraction, Time Dilation, and Relativity of Simultaneity

We have seen how different reference frames are related in special relativity through Lorentz transformations given in (3.5).

Discussion 3.A: **Length Contraction**

Daniel has just purchased a fancy new sports car. When the car is sitting still, he measures it to have a length L.

(a) Now, suppose that Daniel is inside the car, driving it to the right at a velocity u. If he takes a tape measure and measures the length of the moving car that he's in, what does he measure?

(b) Daniel picks his coordinates such that the back of the car sits at $x'_B = 0$. What is the coordinate x'_F of the front of the car?

(c) Anthony is watching Daniel as he drives by. In Anthony's frame, let x_B and x_F be the coordinates of the back and front of the car, respectively. Use the Lorentz transformations to relate x'_F to x_F, and x'_B to x_B.

(d) What is the length of the car that Anthony measures?

(e) Is your answer from part (d) consistent with the length contraction formula?

(f) Bonus: as Daniel drives by Anthony, will Anthony appear wider or thinner than usual?

A consequence of the Lorentz transformations is that different observers measure different distances. Say observer B in the reference frame (t', x'^i) has a ruler with endpoints at $x'_0 = 0$ and $x'_1 = 1$ m. What does observer A see? We can plug these values into (3.5) to see that the endpoints go as:

$$x'_1 = \gamma(x_1 - ut), \qquad x'_0 = \gamma(x_0 - ut), \tag{3.14}$$

so that:

$$\Delta x' = x'_1 - x'_0 = \gamma(x_1 - x_0) = \gamma \Delta x. \tag{3.15}$$

Thus, while B thinks his ruler is $\Delta x' = 1$ m long, A actually sees B's ruler as being $\Delta x = \gamma^{-1}(1m)$ long! Since $\gamma > 1$, A sees the length of B's ruler *contracted*. This is the *length contraction rule: a ruler travelling at a velocity u (with respect to the observer) has its length contracted by a factor of γ^{-1}*.

We can graphically represent length contraction with a spacetime diagram, see Fig. 3.5.

Fig. 3.5 Length contraction in a spacetime diagram. The moving ruler has length l' in the moving reference frame (ct', x') but length $l < l'$ in the reference frame (ct, x)

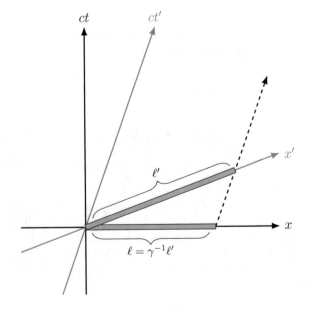

Exercise 3.10 The above discussion told us that observer A sees observer B's ruler's length contracted by a factor of γ. What if observer A had a ruler of 1 m in his reference frame—what would the length be that observer B measures? (You should not need to do any calculation!)

Exercise 3.11 What happens if the ruler is pointing in the y direction, i.e. $y_0 = 0$ and $y_1 = 1$ m (while the relative motion of observers in still in the x-direction); what are $y_0'(t)$, $y_1'(t)$?

Discussion 3.B: Time Dilation

Daniel enters a race! The race track is a straight line, and Daniel is moving to the right at a constant velocity u.

(a) When the race starts, Anthony's stopwatch reads $t_S = 0$. When Daniel finishes the race, Anthony's stopwatch reads that it took him a time t_F to finish. In Anthony's frame, how far did Daniel go?

(b) Anthony has been watching the race from the starting line, so he picks his coordinates such that the starting line is at $x_S = 0$. What is the coordinate x_F of the finish line?

(c) When the race starts, Daniel's stopwatch reads $t_S' = 0$. According to his stopwatch, it takes him a time t_F' to finish the race. Use the Lorentz transformations to relate t_F' to t_F and x_F.

(d) Plug in your answer from part (b) for x_F and try to simplify your answer to part (c). Is your result consistent with the time dilation formula?

Another consequence of (3.5) is that different observers measure different times, as well. If observer B (in frame (t', x'^i)) has a clock at $x_0' = x_1' = 0$ in his reference frame that measures $\Delta t' = t_1' - t_0'$, then observer A will measure:

$$t_0' = \gamma \left(t_0 - \frac{u}{c^2} x_0 \right),$$

$$t_1' = \gamma \left(t_1 - \frac{u}{c^2} x_1 \right) = \gamma \left(t_1 - \frac{u}{c^2} [x_0 + u (t_1 - t_0)] \right) = \gamma^{-1} (t_1 - t_0) + t_0', \quad (3.16)$$

where we used that $x_1' = x_0' + ut$ as the primed reference frame is moving at constant velocity with respect to the unprimed reference frame. We conclude that:

$$\Delta t = \gamma \Delta t'. \quad (3.17)$$

Since $\gamma > 1$, this is the "opposite" kind of effect than we saw with lengths: observer A actually sees observer B's clock ticking slower than B sees. This is the *time dilation rule: a clock travelling at a velocity u (with respect to the observer) is seen to experience time slowed down by a factor γ.*

Fig. 3.6 Time dilation in a
spacetime diagram. A clock
in the moving reference
frame (ct', x') is observed to
take longer (i.e. go slower)
than in the reference frame
(ct, x)

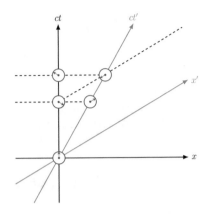

We can also depict time dilation graphically in a spacetime diagram, see Fig. 3.6.

Exercise 3.12   The above discussion holds for a clock that is moving in B's
reference frame as measured in A's frame. What if observer A had a clock in
his reference frame? How would B see time ticking by on this clock? (If this
seems weird, wait for the discussion on the twin paradox coming up!)

Exercise 3.13   The direction of the ruler (aligned with the direction of the
constant velocity or not) mattered for length contraction. Is there any similar
phenomenon for time dilation?

Exercise 3.14   I am traveling at $u = 0.9c$ past you. I pass you at your time
t_0 in the origin. You then see me travelling a distance of one light-year (the
distance light travels in one year) to $x_1 = 1 \, ly$; I arrive at t_1.

(a) How long do you think it takes me to travel one light-year, i.e. what is t_1
    (in your time)?
(b) How long did *I* think it took me to travel to this point? (Hint: do *not* simply
    use the time dilation formula for this!)

If length contraction and time dilation are not mind-boggling enough, another
important consequence of the Lorentz transformations (3.5) is the *relativity of simultaneity: two different observers do not always agree on which events happen at the
same time!* Another equivalent way to say this is that *different observers may disagree
on which event happens before the other.*

Exercise 3.15 Let's illustrate the relativity of simultaneity with an example. Observer A in frame (t, x^i) sees two events happen simultaneously at $t = 0$: event I happens at $x_I = -1\,ly$ (light-year) and event II happens at $x_{II} = +1\,ly$. Observer B is travelling in frame (t', x'^i) at velocity u with respect to A according to (3.5); and observer C is travelling in a frame (t'', x''^i) which is travelling with velocity $-u$ with respect to A. Draw a spacetime diagram with the different rest frames and label the relevant points.

(a) We say observer A "sees" events I and II happen at $t = 0$, but if the events actually happen at $t = 0$, at what time will A actually *see* them happen (i.e. when will he be able to receive news of the event)?

(b) Find the analogy of (3.5) relating observer C's coordinates (t'', x''^i) to observer A's coordinates (t, x^i).

(c) Find t'_I, t'_{II} and t''_I, t''_{II}, the times at which events I and II happen according to observer B and C. Which happens first according to which observer?

It may seem extremely confusing and counterintuitive that a sequence of events may be different for different observers. *What happens to cause and effect?!* We will return to this question later in the discussion of *timelike* and *spacelike* separated points in spacetime.

3.2.5 Apparent Paradoxes

Length contraction, time dilation, and the relativity of simultaneity are three consequences of the Lorentz transformation formula (3.5). These properties are so weird and counter-intuitive with respect to our daily life that it is very easy to get confused by these properties. Since these properties are so counter-intuitive, it is also easier to derive incorrect consequences and effects by improperly using the Lorentz transformation formula (3.5). This has historically led to a few famous apparent "paradoxes" in special relativity. We will see that these are not really paradoxes when we carefully and correctly apply the Lorentz transformation rules to obtain the correct physical description of the problem. If it helps, try to draw the relevant spacetime diagrams for each of the situations in the paradoxes to visualize the solution!

The twin paradox deals with a paradox coming from the principle of time dilation.

Discussion 3.C: The Twin Paradox

The twin paradox is perhaps the most famous paradox in special relativity. Consider a pair of twins on earth; they are exactly the same age. One of the twins boards a spaceship and travels (as seen on earth) at $0.9c$ away from the earth for thirty years. He then turns around and travels back to earth (again, as seen on earth) at $0.9c$ back to earth for another thirty years. For $v = 0.9c$, we have $\gamma = 2.3$.

The twin on Earth has thus aged 60 y (sixty years). He knows special relativity, and thus calculated that the twin that was travelling experienced *time dilation*, that is, his clock ticked slower and thus he has aged less: he calculates that his travelling twin has aged only $60/2.3 = 26$ y. Thus, the twin on earth concludes that his travelling twin will be *younger* than he is.

However, the travelling twin also knows special relativity. If he is travelling at speed $0.9c$ with respect to the earth, then the earth is travelling at speed $0.9c$ with respect to him! Thus, he can perform exactly the same calculation and finds that the twin on earth will be younger than he is!

How can this be possible? Both twins conclude that the other twin must be younger than he is. When the travelling twin returns to earth and gets out of his space ship, which twin will be the older one?

The ladder paradox is an apparent paradox using the principle of length contraction.

Discussion 3.D: **The Ladder Paradox**

Suppose you have a ladder of length 2 m. You want to fit this ladder horizontally (parallel to the ground or x-axis) into your garage, but the garage is only 1 m long. Your garage has a front and back door, and all you want to do (for some reason) is fit the ladder in for a brief moment.

You know special relativity and in particular the principle of length contraction, so you come up with the following way to get the ladder in the garage (for a very small moment). You have the ladder travel at $0.9c$ through the garage; because the ladder's length will be contracted with a factor $\gamma = 2.3$, from your point of view the ladder will actually only measure $2m/2.3 = 0.87$ m long, and so will fit in the garage. When the ladder is completely inside the garage, you briefly close the front and back door at the same time, thereby trapping the ladder in the garage! You then open up the back door again so the ladder can continue on through the garage, as you don't want your ladder to crash against the garage.

From your point of view, the ladder thus fits into the garage due to its length being contracted. However, from the ladder's point of view, it is still 2 m long, and to make matters worse, the garage is now length contracted and so only measures $1m/2.3 = 0.43$ m; thus, from the ladder's point of view, it can certainly not fit inside the garage!

Who is correct, you or the ladder? Will you need to buy a new ladder, or will you simply be able to retrieve it wherever it lands after passing through your garage?

The lighthouse paradox has many guises and appears to violate the principle of not being able to travel faster than light.

Discussion 3.E: The Lighthouse Paradox

A lighthouse has a very powerful beam of light that is able to illuminate very far away objects.The beam of light from the lighthouse rotates at a high speed (high, but lower than c!) around the lighthouse. Let's say the beam of the lighthouse takes $1\,s$ to rotate $180°$ (or $\pi\,rad$). If we have two objects that are two light-years apart (with the lighthouse in the middle), these will be illuminated by this light beam only one second apart. Thus, the end of the light beam is clearly travelling (much) *faster than light*! How is this possible?

There are other, similar formulations of this type of paradox (with similar resolutions). One can also imagine pointing a very powerful laser from the earth to the moon. Then, assuming this laser is hand-held, with a very fast motion of my wrist, I can trace out a *huge* circle on the moon with my laser—if the moon is far enough away, and I make my circle large enough, the laser point on the moon tracing out the circle will clearly be travelling faster than the speed of light.

A third formulation of this paradox involves two sticks in the plane. Let's say one stick is along the x-axis, and the second stick is given by the equation $y = x/2$. Give this second, slanted stick a uniform velocity of $0.9c$ pointing downwards in the y-direction. At point $t = 0$, the sticks intersect at the origin. The intersection point will move to to the right as time goes by, and it is easy to see this point actually moves faster than the speed of light. (Exactly how fast is it moving for this particular setup?)

The scissors paradox is a variation of the lighthouse paradox.

Discussion 3.F: The Scissors Paradox

Imagine you have an enormous pair of scissors where the blades are one light-year long, but the handle is normal-sized. You also have an enormous piece of paper that you put inside the scissors' blades. If you then quickly close the scissors with your hand, you will also be closing the enormous blades quickly. There are now two things travelling faster than light: the cutting point where the blades meets the paper, and the blades of the scissors themselves. Convince yourself that the resolution of the paradox of at least one of these two things travelling faster than light is different than the resolution of the lighthouse paradox above.

Exercise 3.16** What is *wrong* with the following reasoning? A (in frame (t, x^i)) has a ruler in his reference frame with endpoints $x_0 = 0$ and $x_1 = 1$ m. To find out the length of this ruler that B (in frame (t', x'^i)) measures, we plug this into (3.5), to get:

$$x_1' = \gamma(x_1 - ut), \qquad x_0' = \gamma(x_0 - ut), \qquad (3.18)$$

thus, I conclude that $\Delta x' = x_1' - x_0' = \gamma \Delta x$. Since $\gamma > 1$, I conclude that observer B in frame (t', x'^i) sees a *larger* length!

3.3   Minkowski Spacetime

It is clear that time in special relativity is no longer the absolute, universal of Newtonian mechanics. Special relativity tells us that time and space are inexorably intertwined into one encompassing concept of *spacetime*. We will explore the mathematical framework for understanding spacetime, which leads us naturally to concepts of distances and metrics. Let's first discuss these concepts in the familiar, three-dimensional spatial world that Newtonian mechanics lives in, before turning to the four-dimensional spacetime of special relativity.

3.3.1   Distances and Metrics in Cartesian Space; Einstein Summation

The distance between two points $p_1(x_1, y_1)$ and $p_2(x_2, y_2)$ on the plane is given in Cartesian coordinates by:

$$d(p_1, p_2) = \sqrt{(x_2 - x_1)^2 + (y_2 - y_1)^2}. \qquad (3.19)$$

When the two points p_1, p_2 are *infinitesimally* close to each other, we can write:

$$x_2 = x_1 + dx, \qquad (3.20)$$
$$y_2 = y_1 + dy, \qquad (3.21)$$

where dx, dy are infinitesimals. We often call the square of the distance formula for such infinitesimally close points ds^2; see also Fig. 3.7. We obtain an expression for

Fig. 3.7 ds is the (infinitesimal) distance between two infinitesimally close points (x, y) and $(x + dx, y + dy)$

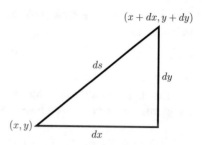

ds^2 by Taylor expanding the distance formula to second order in dx, dy, which is very easy for Cartesian coordinates:

$$ds^2 \equiv d(p_1, p_1 + (dx, dy))^2 = (x_2 - x_1)^2 + (y_2 - y_1)^2 = dx^2 + dy^2. \quad (3.22)$$

We also call ds^2 *the metric* (also sometimes called the *line element*) associated with the Cartesian coordinates (x, y) on the plane.

The metric can be written in numbered coordinate notation as:

$$ds^2 = \sum_{i,j} g_{ij} dx^i dx^j. \quad (3.23)$$

The matrix g_{ij} is called the *metric tensor* (or also sometimes just called the *metric*). For the Cartesian coordinates (x, y), g_{ij} is simply the 2×2 identity matrix. We will see later that the metric can take more complicated forms for other coordinate systems.

The metric tensor can be used to define a *coordinate-invariant* way of defining the dot product:

$$\vec{v} \cdot \vec{w} = \sum_{i,j} v^i g_{ij} w^j. \quad (3.24)$$

We can also use the metric to define a downstairs-index version of any vector:

$$v_i \equiv \sum_j g_{ij} v^j. \quad (3.25)$$

The metric is an invertible matrix, so we can define its inverse:

$$g^{ij} = (g^{-1})^{ij}, \quad (3.26)$$

such that (by definition of matrix inverse):

$$\sum_j g_{ij} g^{jk} = \delta_i^k, \quad \sum_j g^{ij} g_{jk} = \delta_k^i. \quad (3.27)$$

Just as we use the metric to lower indices, we can use the inverse metric to raise an index, so e.g:

$$v^i = \sum_j g^{ij} v_j. \tag{3.28}$$

The metric is a tensor with two covariant (downstairs) indices—this means it transforms under coordinate transformations as in (2.38).

Exercise 3.17 Use your favorite coordinate transformation and convince yourself that g_{ij} indeed transforms as a tensor with two covariant indices.

Exercise 3.18 Prove that the metric for each of the coordinate transformations in Exercise 2.10 is given by (3.22), by using the covariant transformation property in (2.38) of the metric.

Exercise 3.19 Convince yourself that v_i is a covariant tensor, i.e. it transforms as one. Convince yourself that $\sum_{ij} v^i g_{ij} w^j = \sum_i v_i w^i$ transforms as a *scalar*, i.e. as (2.30).

Einstein summation notation is the agreement that indices that appear both "upstairs" and "downstairs" in an expression should be summed over. This allows us to get rid of cumbersome sums in our notation. Such summed-over indices are sometimes called "dummy indices", while non-summed over indices are called "free indices". Summed-over indices are said to be "contracted". For example:

$$v_i = g_{ij} v^j = \sum_j g_{ij} v^j = g_{i1} v^1 + g_{i2} v^2. \tag{3.29}$$

In this equation, j is a dummy index, while i is a free index. The equation should be seen as a vector equation; it has as many equations as i can take on values. The free indices on all terms of an expression should match and should be in the same place (up- or downstairs), otherwise the expression does not make any sense! Dummy indices must appear upstairs and downstairs *exactly once*.

Exercise 3.20 Which of the following expressions are allowed, and which are not?

(a) $T^i_{\ j} v^j = v^i$

(b) $T_{ij} v^j = v^i$

(c) $M^k_{\ k} v^i = v^i$

(d) $M^i_{\ j} v^j \ \mathfrak{I} \ M^k_{\ j} v^j$

(e) $T^i_{\ j} v^j w_j$

Exercise 3.21 Write out the following explicitly in terms of every individual component, i.e. performing the implicit sums:

(a) $v^i v_i$

(b) $T^{ij} v_i v_j$

(c) $T^{ij} v_j$

Do you begin to see the power of the ~~Dark Side~~ Einstein summation convention?

Exercise 3.22 Does the following equation contain any non-trivial information?

$$T_{ij} v^i w^j = T_{kl} v^k w^l. \tag{3.30}$$

Exercise 3.23 What kind of a tensor (scalar, vector, etc.) are the following, i.e. how do they transform under coordinate transformations? How many components do these tensors have? How many *independent* components?

(a) $v^i w_i$

(b) $v^i w_j$

(c) $v^i w^j$

(d) $v^i v^j$

With Einstein summation notation, it becomes very clear how an expression involving multiple indices and their contractions will transform under coordinate transformations: any free contravariant (upstairs) index i will transform as (2.36) while any free covariant (downstairs) index j will transform as (2.38). A corollary is that if an expression has no free indices, it is a scalar!

Exercise 3.24 Let's try to understand the metric a bit better.

(a) Take two different coordinate systems (x, y) and (x', y'), related by $(x' = \lambda x, \, y' = y)$. Try to argue what (3.19) should be in terms of the coordinates x', y', if the distance between two points is to remain the same in the (x', y') and (x, y) coordinate systems.

(b) Use this distance formula in the (x', y') coordinates to find the analogue of the metric in (3.22) in terms of dx', dy', and thus find the metric tensor $g'_{ij}(x'^i)$.

(c) Take a vector with components v^i in the (x, y) coordinate system. Argue what the (same!) vector's components v'^i in the (x', y') should be. (If necessary, try drawing a few simple examples to visualize the problem.)

(d) Now calculate the size of the vector $v^2 = g_{ij} v^i v^j$ in the (x, y) coordinate system, and $v'^2 = g'_{ij} v'^i v'^j$ in the (x', y') coordinate systems. Is the result what you would have expected? Do you see why the metric a useful object?

Fig. 3.8 ds is the (infinitesimal) distance between two infinitesimally close points (r, θ) and $(r + dr, \theta + d\theta)$

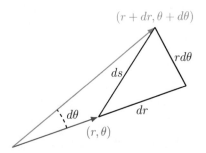

If we have a surface *inside* of space, then we can find the metric on that surface if we have a parametrization of that surface. A simple example makes a lot clear: consider a circle of radius R in the plane. A good parametrization for this is given by:

$$x = R \cos t, \qquad y = R \sin t, \tag{3.31}$$

where the parameter t runs from 0 to 2π. To find the metric ds^2_{circle} on the circle in terms of the infinitesimal change in coordinate t, we simply plug in the parametrization into the metric of the plane, taking appropriate derivatives:

$$ds^2_{circle} = dx(t)^2 + dy(t)^2 = R^2(-\sin t \, dt)^2 + R^2(\cos t \, dt)^2 = R^2 dt^2. \tag{3.32}$$

The process of finding the metric on a surface *inside* another space is called *pulling back* the metric to the surface.

Exercise 3.25*   Find the metric on a sphere; note that a sphere is a two-dimensional surface so you will need two independent parameters in its parametrization! Hint: remember (2.41).

We can also specify vectors and the metric in curvilinear coordinates; see also Fig. 3.8.

Exercise 3.26   Prove that (3.19) is equivalent to:

$$d(p_1, p_2) = \sqrt{r_1^2 + r_2^2 - 2r_1 r_2 \cos(\theta_2 - \theta_1)}, \tag{3.33}$$

in polar coordinates.

Exercise 3.27*   Prove that the metric that follows from (3.33) is (with $p_1(r, \theta)$ and $p_2(r + dr, \theta + d\theta)$):

$$ds^2 = dr^2 + r^2 d\theta^2. \tag{3.34}$$

Exercise 3.28* Alternatively, convince yourself that we can view $x(r, \theta)$ and $y(r, \theta)$ in (2.40) as a *parametrization* of the entire space; this means we can simply find the metric in terms of r, θ by using the formula for an embedded surface as discussed in the previous section.

Exercise 3.29 Find the differential version of the coordinate transformation (2.40), i.e. find dx, dy in function of $dr, d\theta$. Use this to go prove (3.34), starting from (3.22).

Exercise 3.30 Find the polar coordinate components (v^r, v^θ) for the vector fields in Fig. 2.6.

Exercise 3.31 Use your favorite vector field and find expressions for its components (v^r, v^θ) in polar coordinates (using (2.36)). Compute v_i using the metric (3.34) and compare to the components v^i. Check that $v^2 = v^i v_i$ gives the correct answer.

Exercise 3.32* For spherical coordinates (r, θ, ϕ) in 3D:

(a) Find the 3D analogue of the distance formula (3.33) in spherical coordinates. (Use (2.41) if you need it.)
(b) Prove that the 3D metric ds^2 in terms of $dr, d\theta, d\phi$ is given by:

$$ds^2 = dr^2 + r^2 \, d\theta^2 + r^2 \sin^2 \theta \, d\phi^2. \tag{3.35}$$

(c) Work out the components (v^r, v^θ, v^ϕ) of a vector v in terms of the components in Cartesian coordinates (v^x, v^y, v^z).
(d) Find the basis vectors $\vec{e}_i$ for spherical coordinates. Check that $g_{ij} = \vec{e}_i \cdot \vec{e}_j$ for spherical coordinates; then prove that this relation is always true in any coordinate system. (See also Exercise 2.20.)
(e) Where are there coordinate singularities in the spherical coordinate system? (See also Exercise 2.25(a).) Take a few interesting-looking vectors and transform them into spherical coordinates; try to understand what happens at the coordinate singularities with the vector components.
(f) Find $\vec{\nabla}$ in spherical coordinates.

3.3.2 The Spacetime Metric

Remember that the metric for space in Newtonian physics was given by:

$$ds^2 = dx^2 + dy^2 + dz^2. \tag{3.36}$$

Inertial observers are precisely the observers that see the same metric. Remembering that the metric is equivalent to the distance formula telling us the distance between two points x^i and $x^i + dx^i$, we can also say that all inertial observers measure the same distance between two points.

We have seen above that this is no longer true in special relativity! Different inertial observers will observe different distances between two points. To make matters worse, two observers will not even agree anymore on the *time* difference between two events! Clearly, the Newtonian metric (3.36) is not an invariant quantity for different inertial observers.

The invariant metric for special relativity is called the Minkowski spacetime metric and is given by:

$$ds^2 = -c^2 dt^2 + dx^2 + dy^2 + dz^2. \tag{3.37}$$

Exercise 3.33   Prove that (3.37) is indeed invariant under Lorentz transformations, in other words $ds^2(dt, dx, dy, dx)$ and the transformed metric $ds^2(dt', dx', dy', dz')$ have the same form as given in (3.37).

It is clear that it does not make any sense to disentangle time from the other spatial coordinates; it does not make sense to speak of "space" separate from "time": the notion of one depends on the notion of the other, and what we mean by one changes for different inertial observers. Thus, when we speak of coordinates of a point, it makes the most sense to speak of its *spacetime* coordinates. If we denote $(x, y, z) = (x^1, x^2, x^3)$ as usual, and we define:

$$x^0 = ct, \tag{3.38}$$

then the spacetime coordinates of a point are given by:

$$x^\mu = (x^0, x^1, x^2, x^3) = (x^0, x, y, z) = (x^0, x^i) = (x^0, \vec{x}). \tag{3.39}$$

We will use Greek indices μ, ν, ρ, σ etc. to indicate that we are using 4-dimensional spacetime coordinates (as opposed to the 3D spatial i, j indices).

If we write the metric as:

$$ds^2 = g_{\mu\nu} dx^\mu dx^\nu, \tag{3.40}$$

where we sum over μ, ν as always, then we see that:

$$g_{\mu\nu} = \begin{pmatrix} -1 & 0 & 0 & 0 \\ 0 & 1 & 0 & 0 \\ 0 & 0 & 1 & 0 \\ 0 & 0 & 0 & 1 \end{pmatrix}. \tag{3.41}$$

Note that this implies the following relation between x_μ and x^μ:

$$x_0 = -x^0, \quad x_i = +x^i. \tag{3.42}$$

Hopefully, we have learned by now to carefully distinguish between upper and lower indices: in Minkowski spacetime, they are sometimes literally opposites! We note that the inverse metric $g^{\mu\nu}$ also has the same matrix form as $g_{\mu\nu}$.

Exercise 3.34 Replace the 3D Cartesian $i, j, \cdots$ indices by 4D spacetime $\mu, \nu, \cdots$ indices in Exercise 3.21, and redo the exercise.

Exercise 3.35 Consider the following equations; write each component of each equation out explicity, performing all of the implicit sums:
(a) $v_\mu = g_{\mu\nu} v^\nu$
(b) $v^\mu = g^{\mu\nu} v_\nu$
(c) $g^{\mu\nu} g_{\nu\rho} = \delta^\mu_\rho$

3.3.3 Invariant Spacetime Distances: Timelike, Null, and Spacelike

Remember that specifying a metric is equivalent to specifying a distance formula. The distance formula for the spacetime distance between two points $p_1(x_1^\mu)$ and $p_2(x_2^\mu)$ that follows from (3.37) is:

$$d(p_1, p_2)^2 = -(ct_2 - ct_1)^2 + (x_2 - x_1)^2 + (y_2 - y_1)^2 + (z_2 - z_1)^2. \tag{3.43}$$

Lorentz transformations leave this distance invariant: two inertial observers will always agree on the spacetime distance between two points—even though they don't necessarily agree on the time or spatial distance between the two points!

While a distance squared in Newtonian space was always positive, this is not necessarily the case for Minkowski spacetime. We distinguish three possibilities for $d(p_1, p_2)^2$:

- $d(p_1, p_2)^2 > 0$: We say p_1 and p_2 are *spacelike separated*.
- $d(p_1, p_2)^2 < 0$: We say p_1 and p_2 are *timelike separated*.
- $d(p_1, p_2)^2 = 0$: We say p_1 and p_2 are *null separated* (or *lightlike separated*).

Exercise 3.36 Let's try to understand timelike separated points a bit better. Draw a spacetime diagram to help visualize and answer the questions (no calculations needed). It also helps to refer back to Exercise 3.15.

(a) Consider a single observer at the origin. What is $d(p_1, p_2)$ where p_1 and p_2 are the origin at different times?
(b) Prove that for any two timelike separated points in spacetime, there always exists an inertial observer for which the two spacetime points are at the origin but at different times. What does this mean physically? Does it make sense to talk about one spacetime point happening before the other?

Exercise 3.37 Let's try to understand spacelike separated points a bit better. Consider drawing a spacetime diagram to help visualize the questions.

(a) Consider a single observer at the origin. Give two spacetime points p_1, p_2 that are spacelike separated for which $t_2 = t_1$. Also give two spacelike separated points for which $t_2 \neq t_1$.
(b) Prsove that for every two spacelike separated points in spacetime, there exists an inertial observer for which $t_2 > t_1$, another inertial observer for which $t_1 > t_2$, and a third inertial observer for which $t_1 = t_2$. What does this mean physically? Does it make sense in this case to talk about one spacetime point happening before the other?

Exercise 3.38 Consider two null separated points in spacetime for a given observer. Ask yourself the same type of questions that we did for timelike and spacelike separated points.

What we have discovered here is a very deep insight into physics. It does not make sense to talk about a chronological sequence of events unless they are timelike (or null) separated. This is the principle of *causality*: only timelike separated points can influence each other physically. Spacelike separated points are necessarily not in *causal contact* with each other.

Exercise 3.39 Return to the discussion we had previously on the *relativity of simultaneity*. Discuss what the relation is between the distance between events (spacelike or timelike) and whether or not observers can agree on which event happens first.

Exercise 3.40 Argue, only using causality, that physical particles must always travel on *timelike* (or null) curves, i.e. curves for which the distance between two points on the curve are always timelike (or null).

Exercise 3.41* Using the principle of causality, try to understand why Newtonian gravity theory is inconsistent or *acausal*.

3.3.4 Vectors and Tensors

A vector v^μ in Minkowski spacetime now has *four* components since spacetime has four dimensions.

We can again define a downstairs-index version of v^μ using the metric $g_{\mu\nu}$:

$$v_\mu = g_{\mu\nu} v^\nu. \tag{3.44}$$

Note that $v_0 = -v^0$ while $v_i = +v^i$ (similar to $x_0 = -x^0$ and $x_i = +x^i$). We can also define the vector "dot product", given by:

$$v^2 = v^\mu v_\mu = v^0 v_0 + v^i v_i = -(v^0)^2 + (v^1)^2 + (v^2)^2 + (v^3)^2. \tag{3.45}$$

> **Exercise 3.42** Convince yourself that it makes sense to talk about *timelike, spacelike, and null (or lightlike)* vectors, depending on the value of v^2.

Precisely as in Cartesian space, a vector is again a quantity that transforms under coordinate transformations as:

$$v'^\mu = M^\mu{}_\nu v^\nu, \qquad M^\mu{}_\nu = \frac{\partial x'^\mu}{\partial x^\nu}. \tag{3.46}$$

> **Exercise 3.43** Find $M^\mu{}_\nu$ for the Lorentz transformation (3.5). Check that (3.46) gives the right answer for the transformation of the (spatial) velocity given in (3.8).
>
> **Exercise 3.44** We found the rule for addition of velocities above in (3.8). Can this velocity be seen as (the spatial part of) a four-vector in Minkowski space?

We can also define a tensor field similarly, e.g. $T^{\mu\nu}$:

$$T'^{\mu\nu} = M^\mu{}_\rho M^\nu{}_\sigma T^{\rho\sigma}, \tag{3.47}$$

or a tensor with a covariant ("downstairs") index:

$$K'_\mu = (M^{-1})_\mu{}^\nu K_\nu. \tag{3.48}$$

> **Exercise 3.45** Calculate $(M^{-1})_\mu{}^\nu$ for the Lorentz transformation (3.5).

Once again, the "free indices" of an expression tells you how it transforms under Lorentz transformations. For example, the vector dot product $v^2 = v^\mu v_\mu$ is again a scalar.

3.4 Relativistic Mechanics

The velocity $v^i = dx^i/dt$ is no longer a good (four-)vector in Minkowski spacetime (see Exercise 3.44). In essence, this is because both dx^i and dt are components of the same coordinate four-vector $(c\,dt, dx^i)$, and dividing different components of a vector is not a good way to get another vector! The four-vector *proper velocity* u^μ of a particle is defined as:

$$u^\mu = (c\gamma, \gamma\vec{v}) = (c\gamma, \gamma v^x, \gamma v^y, \gamma v^z), \tag{3.49}$$

with:

$$\gamma = \frac{1}{\sqrt{1 - \left(\frac{\vec{v}}{c}\right)^2}}, \qquad \vec{v}^2 = (v^x)^2 + (v^y)^2 + (v^z)^2. \tag{3.50}$$

Exercise 3.46 Check that:

$$u^2 = u_\mu u^\mu = -c^2. \tag{3.51}$$

Exercise 3.47

(a) What is u^μ in the rest-frame of the particle?
(b) Assuming that u^μ is indeed a four-vector, use the Lorentz transformation (3.5) to transform the components of u^μ from the rest-frame to a different frame; do you get (3.49)?

A particle in Newtonian mechanics traces out a line defined by $x^i(t)$, where x^i is the position of the particle in space at each time t. The time t is here a *parameter* and not one of the coordinates. As we know, in special relativity, time has now also been "elevated" to be one of the four coordinates. When we describe the motion of a particle in special relativity, we should then speak of a line $x^\mu(\tau)$ in spacetime, where x^μ are the coordinates (including time) of the particle for the parameter τ. This line in spacetime is called the *worldline* of a particle (see Fig. 3.9), and contains the *entire* information (including history and future) of the motion of the particle. We then have:

$$u^\mu = \frac{dx^\mu}{d\tau}. \tag{3.52}$$

Fig. 3.9 A worldline of a particle in a spacetime diagram. The parameter τ describes the worldline and indicates how much time the particle feels. u^μ is the proper velocity and is the tangent vector along the worldline

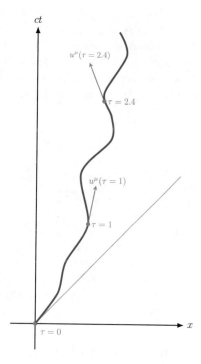

Thus, the four-vector u^μ is the *tangent* vector to the particle's worldline. Since $u^2 = -c^2$, we can write the equation:

$$-c^2 d\tau^2 = -c^2 dt^2 + dx^2 + dy^2 + dz^2. \tag{3.53}$$

The parameter τ is called the *proper time* of the particle: it describes how much time the particle itself experiences *in its own rest frame*.

Exercise 3.48 We can re-derive the time dilation formula using (3.53). Assume the particle (with proper time τ) is travelling at a constant speed in the x-direction, i.e. $u = dx/dt$.

(a) Is the particle moving in the y or z directions? What does this mean for dy and dz in (3.53) for the particle?
(b) Express the particle's dx in terms of its dt, using u.
(c) Now you can use (3.53) to relate the particle's time $d\tau$ to the observer's time dt. Compare the resulting formula with the time dilation formula (3.17).

Exercise 3.49 Cosmic rays are extremely high energy photons that hit the earth's atmosphere and create exotic, heavy particles like *muons*. These particles have an extremely short half-life, but are still able to be detected at the Earth's surface. Let's understand how this works. Assume a muon is created in the atmosphere, 594 km above a detector. It travels at a velocity of $0.99c$ and hits the detector on the ground. The particle has a half-life of about $2.2\,\mu s = 2.2 \cdot 10^{-6}$ s. A particle having half-life $t_{1/2}$ means that if N_0 particles are created at time t_0, after a time t has elapsed, we expect to still have $N(t)$ particles left, where $N(t)$ is given by:

$$N(t) = N_0 \left(\frac{1}{2}\right)^{t/t_{1/2}}. \tag{3.54}$$

The rest of the particle (i.e. $N_0 - N(t)$) have *decayed* into other particles.

(a) How long does it take the muon to hit the detector, from the detector's point of view?
(b) How about from muon's point of view? Explain how relativistic length contraction is necessary for this to make sense.
(c) Calculate $N(t)$ for the muon. Which t do you use (i.e. in which reference frame)? Explain why relativistic time dilation helps us detect more muons created by cosmic rays hitting the atmosphere than would be expected from Newtonian physics.

Using the velocity four-vector, we can define the momentum four-vector of a particle:

$$p^\mu = mu^\mu. \tag{3.55}$$

Let's try to understand some properties of the relativistic momentum better.

Exercise 3.50 Convince yourself (using a Taylor expansion) that when v/c is small, the spatial part of the momentum p^i agrees with what we expect from Newtonian mechanics.

Exercise 3.51 Verify that:

$$p^2 = p^\mu p_\mu = -m^2 c^2. \tag{3.56}$$

We call m the *invariant mass*. What kind of tensor is it (scalar, vector, tensor)?

Exercise 3.52 Now, we turn to the time component of the momentum four-vector, p^0.
(a) Use a Taylor series to expand $p^0 c$ to second order in v/c. Identify the term appearing at second order in v/c. Convince yourself that the logical definition of energy in relativity must be:

$$E_{rel} = p^0 c. \tag{3.57}$$

Also convince yourself that the extra factor of c is necessary.

(b) Now convince yourself that a particle at rest has:

$$E_{rest} = mc^2. \tag{3.58}$$

(c) For a particle with everyday mass travelling at everyday velocities, compare mc^2 and the kinetic energy $1/2m\vec{v}^2$. So, why do we even care about the Newtonian kinetic energy $1/2m\vec{v}^2$?

(d) Nuclear fission is a process where one atom splits into one or more other atoms. In this process, the initial atom has more rest mass m_i than the sum of the final atoms $\sum_f m_f$. For example, when one uranium-235 atom splits, the difference in rest masses of the original is about $\delta m = (m_i - \sum_f m_f) = 3.55 \cdot 10^{-28}$ kg. The mass of one uranium-235 atom is $m_U = 3.9 \cdot 10^{-25}$ kg. What fraction of the uranium atom is converted into energy? (What kind of energy?) How much energy would fissioning one gram of uranium produce? If you know a 100 W light bulb uses 100 J/s, how long can you power such a light bulb with this amount of energy?

Exercise 3.53 A particle has equal rest energy and kinetic energy. How fast is the particle travelling? Compare the answer you get if you consider the Newtonian definition of kinetic energy $E_{kin} = \frac{1}{2}m\vec{v}^2$ to the relativistic kinetic energy $E_{kin} = E_{rel} - E_{rest}$. Use some everyday values for m if need be.

We have seen that relativistic energy is given by:

$$E = p^0 c. \tag{3.59}$$

We can rewrite this as:

$$E^2 = \vec{p}^2 c^2 + m^2 c^4, \tag{3.60}$$

where $\vec{p}^2 = p^i p_i$ denotes the inner product of the spatial momentum $\vec{p}$.

This last relation has the advantage that it can be generalized for a massless particle, such as the photon (denoted by γ):

$$E_\gamma = |\vec{p}|c. \tag{3.61}$$

This equation tells us that a photon also has a relativistic momentum p^μ, with spatial components such that $p^i p_i = E_\gamma^2/c^2$. The photon's four-vector momentum is then:

$$p_\gamma^\mu = \frac{E_\gamma}{c}\left(1, \frac{\vec{v}}{c}\right), \tag{3.62}$$

where $\vec{v}^2 = c^2$.

Exercise 3.54 Let's try to understand this last formula (3.62) better. A photon, i.e. light, must travel at the speed of light. Thus, we must have, assuming the photon travels in the x direction:

$$\frac{dx}{dt} = c. \tag{3.63}$$

(a) In analogy with the massive particle, we can think of the photon as tracing out a worldline given by $x^\mu(\lambda)$, where λ is some parameter telling you where you are on the line. Convince yourself that:

$$0 = -c^2 \left(\frac{dt}{d\lambda}\right)^2 + \left(\frac{dx}{d\lambda}\right)^2 + \left(\frac{dy}{d\lambda}\right)^2 + \left(\frac{dz}{d\lambda}\right)^2, \tag{3.64}$$

for the photon, and that this means that:

$$u^2 = 0, \tag{3.65}$$

for the photon's four-velocity u^μ.

(b) Now, prove that (3.62) is the only possible expression for the photon's four-momentum, if we demand that $p^2 = 0$ and we demand that (3.61) holds.

(c) In particular, does it make sense to speak of "the photon's rest frame"?

Conservation of energy and conservation of momentum can now be summarized as *conservation of total four-momentum* p^μ *in any relativistic system.*

A *force* can be *defined* as something that changes the momentum of a particle (like in Newton's second law):

$$F^\mu = \frac{d}{d\tau} p^\mu(\tau). \tag{3.66}$$

Remember that τ is the parameter along the worldline of the particle, and $u^\mu(\tau)$. The above expression involves the *proper acceleration* four-vector of the particle:

$$a^\mu = \frac{d}{d\tau} u^\mu. \tag{3.67}$$

The four-vector F^μ as defined above is also sometimes called the *Minkowski force* or *proper force*. We can also define the *regular force* $\vec{F}$, given by:

$$\tilde{F}^i = \frac{d}{dt} p^i = \left(\frac{d\tau}{dt}\right) F^i. \tag{3.68}$$

The quantity $\vec{F}$ is not a good four-vector.

Exercise 3.55 Try to understand the acceleration four-vector a^μ a bit better, physically:

(a) Prove that $u \cdot a = u^\mu a_\mu = 0$, i.e. the proper acceleration is always perpendicular to u^μ.
(b) Explain what a^μ is in the frame of the particle itself.
(c) Find $x^\mu(\tau)$ for a particle with constant acceleration in the x-direction, i.e. $a^2 = a^\mu a_\mu = k^2$ for some constant k and $a^y = a^z = 0$. Draw the path of the particle in a spacetime diagram; do you understand why this is called *hyperbolic* motion?

Exercise 3.56 Explain why it does not make sense to have a massless particle in Newtonian mechanics.

Exercise 3.57* *Matter* and *antimatter* annihilate when they come together. An electron with mass m and momentum p collides and annihilates with its antiparticle, a positron, which has the same mass m and is at rest. The annihilation produces two photons. Discuss the momentum and energy of the photons; explain why it is not possible that this process only creates *one* photon.

Exercise 3.58 Explain the difference between *proper* and *regular* force—what kind of observer observes each? Is there a similar kind of distinction in Newtonian physics?

Exercise 3.59* A *tachyon* is a theoretical particle that would travel faster than the speed of light. Assuming $p^2 = -m^2c^2$ still holds, what does this mean for m? What kind of vector is the tachyon's proper four-velocity u^μ? Show that, by considering different possible relativistic reference frames, a tachyon can travel arbitrarily fast, but never *slower* than the speed of light (just as a normal particle can never travel faster).

3.5 Advanced Topics

The topics in this section assume some familiarity with advanced physics and/or math concepts.

3.5.1 Electrodynamics

3.5.1.1 The Players: $\vec{E}, \vec{B}, \rho, \vec{j}$

At the beginning of the chapter, we mentioned that the laws of electrodynamics led to confusion as they did not transform correctly under Newtonian coordinate transformations. Here, we will see that they actually transform naturally under Lorentz transformations.

In electrodynamics, there are *electric and magnetic fields*: these are fields that permeate all of spacetime. Electric fields $\vec{E}$ can be thought of as something indicating the presence of charges—an electric charge is something that, by definition, "radiates" electric field line; see Fig. 3.10. If you are familiar with the electrostatic potential V, the electric field is related to the potential as $\vec{E} = -\vec{\nabla}V$.

Moving charged particle also create magnetic fields $\vec{B}$ (see Fig. 3.11). No magnetically charged particles have ever been observed. We are familiar with magnets: these have a magnetic field which, in a sense, is thought to be generated by continuously moving charged particles within the magnet.

The electric and magnetic fields clearly have a magnitude and direction at every point: they are spatial (!) vector fields E^i, B^i (or $\vec{E}$, $\vec{B}$). We can also denote the electric charge density field as ρ: this describes how much electric charge there is present at every point in space. Finally, we can also describe *moving* charge densities with the current spatial vector field $\vec{j} = \rho\vec{v}$, where $\vec{v}$ is the velocity vector of the charge ρ at every point. We can never create or destroy electric charges; this is codified in the *continuity equation*:

$$\vec{\nabla} \cdot \vec{j} = -\frac{\partial \rho}{\partial t}. \tag{3.69}$$

In essence, this equation is saying that any change in charge density at a given point (the right-hand side of the equation) must be the effect of this charge density moving in a current $\vec{j}$ away from this point. (Compare this to the conservation of energy-momentum (4.84)!)

The quantities $\vec{E}$, $\vec{B}$, ρ, $\vec{j}$ are the "players" in electrodynamics. There are also constants ϵ_0, μ_0, which will be mostly unimportant for the concepts we are discussing.

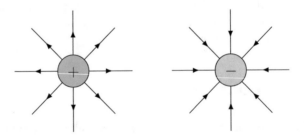

Fig. 3.10 A charged point particle with the electric field as emanating from the particle. The direction of the electric field lines is either outwards (for a positively charged particle) or inwards (for a negatively charged particle); the magnitude of the electric field is larger if the particle's charge is larger

Fig. 3.11 A magnet with magnetic field lines. Note that the magnetic field lines come from the north pole and go to the south pole, and are such that they always form closed loops

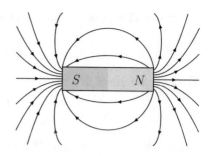

3.5.1.2 The Laws of Electrostatics

In electrostatics, we allow for electric and magnetic fields that are constant in time; the current and charge density must also be constant in time.

Exercise 3.60 Explain how both the current and the charge density could possibly both be constant in time; give some physical examples.

As we said, the electric field indicates the presence of electric charge. This can be codified in *Gauss' law*:

$$\vec{\nabla} \cdot \vec{E} = \frac{1}{\epsilon_0}\rho. \tag{3.70}$$

Mathematically, this equation simply says that "the magnitude of $\vec{E}$ depends on the magnitude of ρ, and the direction of $\vec{E}$ will be *away* from the charge ρ". You may be more familiar with the integral version of the law:

$$\oint_S \vec{E} \cdot d\vec{a} = \frac{1}{\epsilon_0}Q, \tag{3.71}$$

which says that the total "electric flux" (which is basically the integral of $\vec{E}$) through the surface S is equal to the amount of charge enclosed in the surface.

The force exerted on a charged particle due to an electric field is, with Q the charge of the particle:

$$\vec{F}_e = Q\vec{E}. \tag{3.72}$$

Exercise 3.61 Argue that this fits into our intuition that oppositely-charged particles attract and same types of charge repel.

The equation telling us how the magnetic field is related to a current is *Ampère's law*:

$$\vec{\nabla} \times \vec{B} = \mu_0 \vec{j}. \tag{3.73}$$

The integral form of this is:

$$\oint_C \vec{B} \cdot d\vec{l} = \mu_0 I, \tag{3.74}$$

which relates the line integral of $\vec{B}$ along a circle with the total amount of current I passing through that circle.

The force exerted on a charged particle of charge Q and velocity $\vec{v}$ due to a magnetic field is:

$$\vec{F}_m = Q\vec{v} \times \vec{B}, \tag{3.75}$$

thus, the force acting on the particle is always perpendicular to both its velocity and the magnetic field!

The fact that there is no magnetically charged particle is codified in the expression:

$$\vec{\nabla} \cdot \vec{B} = 0. \tag{3.76}$$

We also have the equation:

$$\vec{\nabla} \times \vec{E} = 0. \tag{3.77}$$

3.5.1.3   Electrodynamics and Maxwell's Laws

So far we have always demanded that the magnetic and electric fields are constant in time. However, one can imagine certain experiments of scenarios where this is not the case. Faraday noticed that *a changing magnetic field indices an electric field*, and codified this in *Faraday's law*:

$$\vec{\nabla} \times \vec{E} = -\frac{\partial \vec{B}}{\partial t}. \tag{3.78}$$

So far, the resulting equations and "laws" were the painstaken results of many experiments by different people (hence the different names attached to the laws!) at different times. Maxwell was the first to combine these and notice a "missing piece".

Let's combine the divergenceless of the magnetic field, $\vec{\nabla} \cdot \vec{B}$ with Faraday's law. We can take:

$$\vec{\nabla} \cdot (\vec{\nabla} \times \vec{E}) = \vec{\nabla} \cdot \left(-\frac{\partial \vec{B}}{\partial t}\right) = -\frac{\partial}{\partial t}(\vec{\nabla} \cdot \vec{B}). \tag{3.79}$$

The left-hand side is identically zero because $\vec{\nabla} \cdot (\vec{\nabla} \times \vec{v}) = 0$ for any vector $\vec{v}$. The right-hand side is zero because $\vec{B}$ is divergenceless, so everything is fine.

However, let's do the same for $\vec{B}$, combining Ampère's law with the continuity equation:

$$\vec{\nabla} \cdot (\vec{\nabla} \times \vec{B}) = \mu_0 (\nabla \cdot \vec{j}) = -\mu_0 \frac{\partial \rho}{\partial t}. \tag{3.80}$$

The left-hand side is again identically zero, but what about the right-hand side? We can apply Gauss' law to get:

$$-\mu_0 \frac{\partial \rho}{\partial t} = -\mu_0 \frac{\partial}{\partial t} (\epsilon_0 \vec{\nabla} \cdot \vec{E}) = -\epsilon_0 \mu_0 \vec{\nabla} \cdot \left(\frac{\partial \vec{E}}{\partial t} \right). \tag{3.81}$$

This expression is not necessarily zero, so there is a problem. Maxwell realized this, and proposed to change Ampère's law to:

$$\vec{\nabla} \times \vec{B} = \mu_0 \vec{j} + \mu_0 \epsilon_0 \frac{\partial \vec{E}}{\partial t}. \tag{3.82}$$

Exercise 3.62 Prove that this alteration of Ampère's law indeed solves the problem, so that $\vec{\nabla} \cdot (\vec{\nabla} \times \vec{B}) = 0$, as it should be.

Thus, not only does a changing magnetic field induce an electric field, but a changing electric field also induces a magnetic field. Maxwell's suggestion was later confirmed by Hertz' experiments on electromagnetic waves.

We can summarize Maxwell's laws by:

$$\vec{\nabla} \cdot \vec{E} = \frac{1}{\epsilon_0} \rho, \qquad \vec{\nabla} \times \vec{E} = -\frac{\partial \vec{B}}{\partial t}, \tag{3.83}$$

$$\vec{\nabla} \cdot \vec{B} = 0, \qquad \vec{\nabla} \times \vec{B} = \mu_0 \vec{j} + \mu_0 \epsilon_0 \frac{\partial \vec{E}}{\partial t}. \tag{3.84}$$

Exercise 3.63 (*Electromagnetic duality*) Somehow, Maxwell's equations are not *quite* satisfactory. It seems like there is *almost* a symmetry "exchanging" the electric and magnetic fields.

(a) Introduce a magnetic charge ρ_m. Where would it naturally enter the equations? Don't forget to allow the magnetic charge to have a non-zero velocity!

(b) Prove that now the equations are invariant under the transformation:

$$\vec{E}' = \vec{B}, \qquad \vec{B}' = -\mu_0 \epsilon_0 \vec{E}, \tag{3.85}$$

as long as we also change exchange the electric and magnetic charges (how?). This is called *electromagnetic duality*.

Exercise 3.64 Electromagnetic waves The existence of electromagnetic waves in vacuum ($\rho = 0$, $\vec{j} = 0$), e.g. light, is included in Maxwell's laws.

(a) Using Maxwell's laws, prove that:

$$\vec{\nabla} \times (\vec{\nabla} \times \vec{E}) = -\mu_0 \epsilon_0 \frac{\partial^2 \vec{E}}{\partial t^2}, \tag{3.86}$$

and the same equation for $\vec{B}$.

(b) Use $\vec{\nabla} \times (\vec{\nabla} \times \vec{v}) = \vec{\nabla}(\nabla \cdot \vec{E}) - \vec{\nabla}^2 \vec{E}$ to conclude that:

$$\vec{\nabla}^2 \vec{E} = \mu_0 \epsilon_0 \frac{\partial^2 \vec{E}}{\partial t^2}. \tag{3.87}$$

Note that $\vec{\nabla}^2 = \partial_x^2 + \partial_y^2 + \partial_z^2$. Thus, every component of $\vec{E}$, $\vec{B}$ satisfies a *wave equation*, where the wave travels with velocity $c = \epsilon_0 \mu_0$.

(c) Prove that $A \sin(kz - \omega t)$ is a solution to the wave equation. What is the relation between k and ω?

(d) Take $E_x = A \sin(kz - \omega t)$ and use Maxwell's equations to find $\vec{B}$. In what direction does the wave propagate; how is this related to the directions of $\vec{E}$, $\vec{B}$?

Exercise 3.65 Electrodynamic energy and momentum Use Maxwell's laws to prove the conservation law for energy and momentum of the electromagnetic fields:

$$\frac{\partial u_{em}}{\partial t} = -\vec{\nabla} \cdot \vec{S}, \tag{3.88}$$

where u_{em} is the energy stored in the electromagnetic fields, and is given by:

$$u_{em} = \frac{1}{2} \left(\epsilon_0 \vec{E}^2 + \frac{1}{\mu_0} \vec{B}^2 \right), \tag{3.89}$$

and $\vec{S}$ is the *Poynting vector*, which is the *energy flux density*, or *energy per unit time, per unit area* transported by the electromagnetic fields, given by:

$$\vec{S} = \frac{1}{\mu_0} (\vec{E} \times \vec{B}). \tag{3.90}$$

Electrodynamics in Special Relativity

If electrodynamics and special relativity are indeed both "correct" theories, they should be compatible. Mathematically, this means that it should be possible to express electrodynamics in terms of four-vectors and tensors living in Minkowski spacetime and transforming with Lorentz transformations. It may seem that it is hopeless to consider electric and magnetic fields, being spatial vectors, as objects in Minkowski spacetime. However, we will first state a result and leave it to the exercises for you to convince yourself of its validity.

The electric and magnetic field are part of a *anti-symmetric two-tensor* $F^{\mu\nu}$ in this way:

$$F^{\mu\nu} = \begin{pmatrix} 0 & E^x/c & E^y/c & E^z/c \\ -E^x/c & 0 & B^z & -B^y \\ -E^y/c & -B^z & 0 & B^x \\ -E^z/c & B^y & -B^x & 0 \end{pmatrix}. \tag{3.91}$$

In other words,

$$F^{0i} = -F^{i0} = E^i/c, \qquad F^{ij} = -F^{ji} = B^k, \tag{3.92}$$

where $(ijk) = (xyz)$ and cyclic permutations thereof (i.e. (zxy), (yzx)). There is also the *dual tensor* $G^{\mu\nu}$ (which is also antisymmetric):

$$G^{\mu\nu} = \begin{pmatrix} 0 & B^x & B^y & B^z \\ -B^x & 0 & -E^z/c & E^y/c \\ -B^y & E^z/c & 0 & -E^x/c \\ -B^z & -E^y/c & E^x/c & 0 \end{pmatrix}. \tag{3.93}$$

The charge and current density can be combined in a four-vector:

$$j^\mu = \rho u^\mu, \tag{3.94}$$

where u^μ is the proper velocity of the charge density ρ.

All of Maxwell's equations can now be written in the incredibly elegant and compact form:

$$\partial_\mu F^{\mu\nu} = j^\nu, \qquad \partial_\mu G^{\mu\nu} = 0, \tag{3.95}$$

and the continuity equation is:

$$\partial_\mu j^\mu = 0. \tag{3.96}$$

The four-vector electrodynamic force on a particle of charge Q is given by:

$$F^\mu_{rel} = Q F^{\mu\nu} u_\nu. \tag{3.97}$$

Exercise 3.66 If $F^{\mu\nu}$ indeed transforms as a tensor under Lorentz transformations, find how E^i, B^i transform into each other under Lorentz transformations.

Exercise 3.67*** If we only had knowledge of the electric field and special relativity, we would inexorably be led to the existence of magnetic fields. Let us investigate how. (This exercise is inspired heavily by the discussion in Sect. 12.3.1 in Griffiths' "Introduction to Electrodynamics"; see the Bibliography at the end of the book for the full reference.)

Consider a wire (a line) along the x-axis. This wire has a bunch of positive charges (charge density $+\lambda$) moving the right with speed v, and a bunch of negative charges (charge density $-\lambda$) moving to the left with speed v. Clearly, the amount of positive charge at any point is the same as the amount of negative charge, so the wire is not charged—i.e. $\lambda_{tot} = \lambda - \lambda = 0$. The total current density is then $I = 2\lambda v$, where λ is the charge density of the wire. Consider a point particle a distance s away from the wire, travelling to the right with speed u, with $u < v$.

(a) What is the electric force on this point particle?
(b) Now, consider a Lorentz transformation that takes us to the rest frame of the particle. What is the transformed velocity v'_- of the *negative* particles? And what is the transformed velocity v'_+ of the *positive* particles?
(c) Argue that the Lorentz contraction between positive/negative charges implies that the transformed charge densities for the positive and negative charges are $\lambda'_\pm = \pm\gamma_\pm\lambda_0$, where $\gamma_\pm$ are the γ-factors associated with $v_\pm$ (what is λ_0?). Thus, the total transformed charge density is $\lambda'_{tot} = \lambda'_+ - \lambda'_- = \lambda_0(\gamma_+ - \gamma_-)$; show that this can be written as:

$$\lambda'_{tot} = -2\lambda\frac{uv}{c^2}\frac{1}{\sqrt{1 - u^2/c^2}}. \tag{3.98}$$

(d) The electric force due to a charge density λ'_{tot} at a distance s from the wire (where the particle sits) has a magnitude:

$$E' = \frac{\lambda'_{tot}}{2\pi\epsilon_0 s}. \tag{3.99}$$

What is the direction of $\vec{E}'$? What is the electric force acting on the particle due to $\vec{E}'$? Is there a magnetic force acting on this particle in this frame?
(e) Being careful about what quantity is the one that transforms as a four-vector, find the force on the particle in the original reference frame (where there was no net charge on the wire). Compare to the magnetic force due to the magnetic field of a wire with current I at distance s from the wire, where the field has magnitude:

$$B = \frac{\mu_0 I}{2\pi s}, \tag{3.100}$$

and direction pointing away from the wire.

We have thus shown that in one frame, there is a force acting on the particle due to an electric field. In another frame, there is no net charge and thus no electric field, but there must still be a non-zero force—this is, of course, precisely the force due to the magnetic field of the current on the wire!

Exercise 3.68 Compare (3.97) with (3.72) and (3.75).

Exercise 3.69 ϵ-**tensors** Define a four-tensor $\epsilon_{\mu\nu\rho\sigma}$ in spacetime by giving one component $\epsilon_{0123} = 1$ and demanding that it is *totally antisymmetric*: this means that whenever we swap two adjacent indices, there is a minus sign, i.e.:

$$\epsilon_{\mu\nu\rho\sigma} = -\epsilon_{\nu\mu\rho\sigma} = -\epsilon_{\mu\rho\nu\sigma} = -\epsilon_{\mu\nu\sigma\rho}. \tag{3.101}$$

(a) Convince yourself that this completely determines all components of $\epsilon_{\mu\nu\rho\sigma}$. How many (and which) components are non-zero?
(b) Does this definition of the ϵ-tensor make sense; i.e. is $\epsilon_{\mu\nu\rho\sigma}$ really a tensor, defined in this way?
(c) Prove that:

$$G^{\mu\nu} = \frac{1}{2}\epsilon^{\mu\nu\rho\sigma} F_{\rho\sigma}. \tag{3.102}$$

Note that $\epsilon_{0123} = -\epsilon^{0123}$.
(d) Define a similar epsilon tensor ϵ_{ijk} in three-dimensional space. Prove that:

$$(\vec{A} \times \vec{B})^i = \epsilon^{ijk} A_j B_k. \tag{3.103}$$

3.5.2 Photons as Waves

We saw that a photon's four-momentum is given by:

$$p_\gamma^\mu = \frac{E_\gamma}{c}\left(1, \frac{\vec{v}}{c}\right), \tag{3.104}$$

where $\vec{v}$ indicates the direction of the momentum and $\vec{v}^2 = c^2$.

A photon is a particle of light. Light is a *wave* travelling at velocity c and can be described by a *wavelength* and *frequency*. The relation between velocity c, wavelength λ, and frequency f for a wave is:

$$c = \lambda f. \tag{3.105}$$

Only for a photon (i.e. not for general waves), there is also an additional relation between the energy of a photon and its frequency (or wavelength):

$$E_\gamma = hf,$$ (3.106)

where h is *Planck's constant* and is given by:

$$h = 6.62 \cdot 10^{-34} \, J \cdot s.$$ (3.107)

The relation (3.106) comes from *quantum mechanics*; we will not discuss its origin here.

General Properties of Waves

A wave can be thought of as a periodic disturbance of a medium. (Although, for a photon, we saw that this is not really true—there is no aether!) A *transverse* wave is where the wave itself travels transverse (perpendicular to) the disturbance it creates; a *longitudinal* wave travels along the same direction as the disturbance. The prototype description of a wave along the x-axis is a sinusoid (although the wave need not be described by a pure sine function):

$$f(x) = A \sin(kx - \omega t).$$ (3.108)

Try to visualize this wave; it is moving along the x-axis in the positive direction. The *amplitude* or "strength" of the wave is A. The *wavelength* λ of a wave is its "size", more precisely: it is the length between two peaks of the wave—here, it is:

$$\lambda = \frac{2\pi}{k}.$$ (3.109)

(k is called the *wavenumber* of the wave.) The (linear) *speed* of the wave is the speed with which the peaks of the wave move. Here, we have:

$$v = \frac{\omega}{k}.$$ (3.110)

The *frequency* f of the wave is how many times per second a peak passes through a given point; it is always given by:

$$f = \frac{v}{\lambda},$$ (3.111)

so a waves frequency always *decreases* if its wavelength *increases* (and vice versa) if its speed remains constant.

There are many examples of waves in physics. Sound is a (longitudinal) wave; the speed of sound is a constant in air (of a given density and temperature) while the frequency of sound is related to the pitch we hear. Waves in and on fluids such as water are also types of (transverse) waves, but these typically have more complicated descriptions depending on the particular setup.

A photon's wavelength (or frequency) is related to what kind of electromagnetic wave it is; see Fig. 3.12.

While a photon's *velocity* must always be given by c in any inertial frame in special relativity, its energy and therefore its frequency or wavelength is not invariant.

Exercise 3.70   The relativistic Doppler effect Take a photon moving along x-axis, heading in the positive direction. Now do a Lorentz tranformation (3.5) on its momentum vector p^μ where we move to a frame that is moving with a relative velocity u in the x-direction.

(a) Find the expression for the transformed vector p'^μ. What is the transformed E'_γ in terms of E_γ? What is the transformed frequency f' and wavelength λ' in terms of the original f, λ?

(b) Using the results from the previous question, can you understand why we say the light is "blueshifted" or "redshifted", depending on the sign of u?

(c) Think of a relevant case in astrophysics where this effect might be important!

Blueshifting and redshifting are called the *relativistic Doppler effect*, in analogy with the regular Doppler effect for sound. The changing of the pitch of a siren as it goes past you is precisely due to the Doppler effect for sound.

More Exercises on Special Relativity

Exercise 3.71* Using just the two postulates of special relativity:

(A) The speed of light for all observers is constant (and equal to c),

(B) All inertial frames (or observers) are equivalent, i.e. the laws of physics should be the same in each inertial frame,

we can derive the Lorentz transformations (3.5) relating a coordinate system (t, x) to a coordinate system (t', x') that is moving with respect to the (t, x) coordinate system with constant velocity $u = dx/dt$.

(a) The most general linear relation between the two coordinate systems is:

$$x' = ax + bt, \tag{3.112}$$

$$t' = dx + et, \tag{3.113}$$

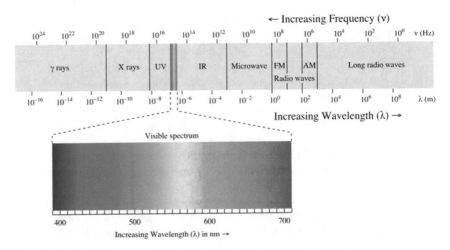

Fig. 3.12 The wavelength spectrum of electromagnetic waves ("EM Spectrum" by Philip Ronan is licensed under CC BY 2.0.)

where a, b, d, e are constants (that may depend on u and c). Why do we want a linear transformation between the two coordinate systems? (Use postulate (B).)

(b) Use postulate (A) to get a relation that the constants a, b, d, e need to satisfy.

(c) The origin $x' = 0$ in the (t', x') frame is moving at a velocity $u = dx/dt$ in the x-direction relative to the (t, x) frame; use this to get a relation between a and b.

(d) How is the origin $x = 0$ in the (t, x) frame moving relative to the (t', x') frame? Use postulate (B) and the *inverse* of the linear transformations (3.112)–(3.113) to get a relation between b and e.

(e) Now set $a = \gamma$ and solve the three relations between a, b, d, e that you found previously. You should recover the correct form of the Lorentz transformations!

(f) Note that we have not determined the correct value of γ yet. To determine γ, demand that $x \cdot x' = x(t', x') \cdot x'(t, x)$; this should lead to the correct expression for γ. (Alternatively, you can demand that the determinant of the coordinate transformation matrix is 1.)

Exercise 3.72 Physicists often use "natural units" instead of SI units where $c = 1$. This means we measure distances *and* times in m. What does a time difference of $\Delta t = 1$ m mean physically in this case? (Hint: Think of the distance travelled by light.) Going through what we have learned, try to understand why we call these units "natural", at least from the perspective of special relativity. If you have an expression given to you in natural units, how do you convert it back into SI units?

Exercise 3.73** Show that it is possible to outrun a light ray if a constant force is applied to you. Do you need a head start?

Exercise 3.74* Explain the (non)sense of the following old limerick:

There was once a girl named Ms. Bright,
Who could travel much faster than light.
She departed one day,
The Einsteinian way,
And returned on the previous night.

Exercise 3.75**

A common misconception is that general relativity is needed to understand the precise resolution of the twin paradox. In this exercise, we will perform an exact computation (in special relativity), determining in a specific situation all quantities completely. Consider that the first of the twins, Anthony, stays at home on Earth in (inertial) reference frame (t, x). The second of the twins, Zach, leaves Earth at $t = 0$ and accelerates away from Earth for a while, then reverses the direction of his acceleration until he accelerates home; finally, he decelerates again. Zach's velocity (as seen by Anthony) is such that he is at an exact standstill when he leaves and arrives on Earth. Let's say his velocity is given by (with k some constant)

$$v = \frac{dx}{dt} = kt \qquad\qquad 0 \le t \le t_1, \qquad (3.114)$$

$$v = \frac{dx}{dt} = kt_1 - k(t - t_1), \qquad\qquad t_1 \le t \le 3t_1, \qquad (3.115)$$

$$v = \frac{dx}{dt} = -kt_1 + k(t - 3t_1), \qquad\qquad 3t_1 \le t \le 4t_1. \qquad (3.116)$$

(a) Draw a diagram of Zach's worldline; convince yourself that his velocity is smooth, i.e. does not make any discontinuous jumps.
(b) What is the maximum velocity v_* Zach achieves (in either direction)? What restrictions must we put on the parameters k and t_1 for consistency with special relativity?
(c) What are the dimensions (length, time, etc.) of k? Calculate d^2x/dt^2 for the entire worldline; what is this quantity physically?
(d) Use the equation relating Zach's proper time differential $d\tau$ to Anthony's dt, dx to find $d\tau/dt$. Be careful with taking square roots!
(e) Use the following equation to find $\tau(t)$ along the entire worldline:

$$\frac{d\tau}{dt} = \sqrt{1 - (at + b)^2} \quad\leftrightarrow\quad \tau(t) = \frac{1}{2a}\left((at + b)\sqrt{1 - (at + b)^2} + \arcsin(at + b)\right) + const.$$
$$(3.117)$$

Do you see from the formula for $\tau(t)$ why the restriction we found above on k, t_1 (or equivalently on v_*) is necessary for things to make sense?

(f) Now find $dx/d\tau$ (still in function of t) by using the chain rule for differentials; you now have the components of the proper velocity $u^\mu(t)$—check that $u^2 = -c^2$.

(g) Again, using the chain rule, calculate $d^2x/d\tau^2$ and thus the proper force needed on Zach along his worldline at every point. How does the force evolve with time (when is it largest)?

(h) Finally, what we came here for! Zach arrives back on Earth after a period of time (as measured by Anthony) of $4t_1$. How much time has passed for Zach? (Here it will be extra important that you were careful with taking square roots to calculate $d\tau/dt$; note that τ should always be increasing!) Express your answer in terms of $\tilde{v}_* = v_*/c$ and t_1. Now, consider the ratio of elapsed time for Zach to elapsed time for Anthony, and see that this is:

$$(ratio) = \frac{\tilde{v}_*\sqrt{1 - \tilde{v}_*^2} + \sin^{-1}\tilde{v}_*}{2\tilde{v}_*}. \qquad (3.118)$$

Take $\tilde{v}_* = 0, 1$ to convince yourself that the endpoints of this function of $\tilde{v}_*$ make sense. Then take some intermediate points (e.g. $\tilde{v}_* = \sqrt{2}/2$ and convince yourself that this ratio is always smaller than one—thus, Anthony has felt *less* time go by than Zach has!

Exercise 3.76   The metric of spacetime in special relativity is (3.37). This follows what is called the "mostly plus" conventions. We could also have defined the metric to be:

$$ds^2_{MM} = c^2dt^2 - dx^2 - dy^2 - dz^2, \qquad (3.119)$$

following the "mostly minus" convention. Go through the chapter; what changes? Try to see that the physics in mostly minus conventions is the same.

Exercise 3.77   Prove that Lorentz tranformations, together with (spatial) rotations, form a *group* by checking the group axioms (closure, associativity, identity, inverse). This group is called the *Lorentz group*. Convince yourself that the Lorentz group, together with translations, also forms a group—this is called the *Poincaré group*.

Chapter 4
General Relativity

4.1 Principles of General Relativity

Before diving in to the mathematics of general relativity, it is important to understand what the physical, conceptual motivations were to consider such a radically new theory of gravity. In this section, we will discover the physical principles underlying general relativity, including the most important one of all: the equivalence principle.

4.1.1 Newtonian Gravitation and Different Types of Mass

> **Discussion 4.A: Different Masses**
>
> Let's think carefully about the concept of "mass", and its physical meaning in the places it appears in Newtonian mechanics:
>
> (a) $F = ma$: What is m here? If I push something with a force F, what does a larger or smaller m mean?
> (b) $F = mg$ (where $g = G\frac{M}{R^2}$): What is m here? If g is the same, what does larger or smaller m mean?
> (c) Are the two m's above the same? Why?! Is there a reason for this in Newtonian mechanics?
> (d) Extra: What about M? This is also a mass; does this play the same role as in (a) and (b)?

© Springer Nature Switzerland AG 2019
D. R. Mayerson et al., *Relativity: A Journey Through Warped Space and Time*,
https://doi.org/10.1007/978-3-030-18914-3_4

Discussion 4.B: The Equivalence Principle

You are in an elevator with no windows or any contact with the outside of the elevator; the elevator is also not tethered or connected to anything, no forces except perhaps gravity can work on it. Consider the following three different possibilities:

(a) The elevator is "floating", i.e. in empty space; there are no forces at all working on it.
(b) The elevator is sitting or "falling" in a universal gravitational field with strength $\vec{g}$ (pointing downwards).
(c) The elevator is sitting or "falling" in the gravitational field of the Earth (where the Earth is located below you).

What is the difference between these three setups? Devise experiments that can determine in which situation your elevator is in. (Use balls as experimental tools if you want.)

In the chapter on classical mechanics, we introduced the Newtonian gravitational force acting on a particle of mass m:

$$F_{grav} = m\,g, \tag{4.1}$$

where g depends on what is responsible for the gravitational force and how far away it is ($g = GM/R^2$). We also introduced Newton's third law, which told us how the particle of mass m responds to a force applied to it by accelerating:

$$F = m\,a. \tag{4.2}$$

But wait! These are two *entirely different* phenomena, but yet they involve the same parameter m! Equation (4.1) tells us what the force on the particle is due to an external gravitational field g, and (4.2) tells us how a particle *responds* to any force by accelerating. In Newtonian physics, they could certainly be different: we could have a *gravitational mass* m_{grav}, such that:

$$F_{grav} = m_{grav}\,g, \tag{4.3}$$

and an *inertial mass* m_{inert}, such that:

$$F = m_{inert}\,a. \tag{4.4}$$

Then, a particle in a gravitational field would accelerate according to:

$$a = \frac{m_{grav}}{m_{inert}}\,g = k\,g. \tag{4.5}$$

The (dimensionless) ratio k could, in principle, be *different* for different kinds of particles or matter; there is absolutely no principle in Newtonian mechanics that dictates that k *must be* some kind of universal constant, the same for all matter. However, as we have silently assumed—and experiments have verified—k *is indeed a universal constant*; in other words, *gravitational mass and inertial mass are the same!*

Exercise 4.1 There are actually *three* types of mass, because in principle the mass that a particle uses to *pull* in the gravitational force m_{pull} can be different than the mass that the particle uses to *feel the pull* m_{grav} due to a gravitational field. The force *on* particle B *due to* particle A is then:

$$F_{grav} = \frac{G}{R^2} m_{pull}^A m_{grav}^B, \tag{4.6}$$

whereas the force *on* particle A *due to* particle B is then:

$$F_{grav} = \frac{G}{R^2} m_{pull}^B m_{grav}^A. \tag{4.7}$$

Convince yourself that equal action and reaction (Newton's third law) assures us that $m_{pull} = k' m_{grav}$ for any kind of matter with necessarily the *same* constant k'—therefore we can simply set $m_{pull} = m_{grav}$ for all matter by "absorbing" the constant k' into the constant G.

Newtonian physics does not provide any explanation for this equality of the different types of masses. Once we set $m_{grav} = m_{inert}$, notice that now we find the extremely simple equation relating a particle's acceleration a to the gravitational field g:

$$a = g. \tag{4.8}$$

This equation does not depend at all anymore on the mass of the particle! Said differently, all objects fall at the same speed, regardless of their mass. This was noticed in the famous experiment by Galileo, and later spectacularly confirmed by astronauts on the moon (where there is no air resistance) dropping a feather and a heavy object and seeing them fall at the exact same rate.

This same principle is often said to be one of Einstein's major inspirations in formulating general relativity, usually formulated using the following thought experiment with elevators. Consider two observers, each one in a sealed off elevator (or other type of box). One elevator is at rest in empty space with nothing around it, and the other one is falling down—towards Earth, for example—unhindered by any cables or anything else stopping it from falling. (We assume the Earth or whatever is pulling the second elevator uniformly pulls at all parts of the elevator.) Now, Einstein's inspiration was the following: *there is no way to distinguish between the two elevators*; in other words, there is no experiment the observers in either elevator could perform to distinguish in which of the two elevators (at rest, or uniformly falling in a gravitational field) they are sitting.

Exercise 4.2 Is the same true of an elevator sitting in an electric and/or magnetic field? Can you think of an experiment to distinguish this (think of redoing Discussion 4.B where you substitute gravitational field with electric/magnetic field)? Why is $m_{grav} = m_{inert}$ crucial for this (thought) experiment?

We have discovered the *principle of equivalence* (again!). This time, the equivalence is not only between different inertial observers, but *uniformly accelerating observers*. Another way of saying this is that all observers are equivalent if there are no forces *other than the gravitational force* working on the observer. To distinguish this version of the principle of equivalence from the others we have previously discussed, this one is sometimes called the *Einstein principle of equivalence*. This principle of equivalence is much deeper and far-reaching than what we previously discussed: it clearly tells us that the gravitational force is *different* than other forces—the source of this fundamental difference is clearly the fundamental principle following from $m_{grav} = m_{inert}$.

The "No Accidents Happen" Principle in Physics

The astute reader might wonder why we are making a big deal out of the equality of m_{grav} and m_{inert} for all types of matter. After all, Newtonian mechanics and gravitational theory are certainly compatible with simply setting $m_{grav} = m_{inert}$ in all Newtonian physics formulae—so is there even a new theory necessary?

This touches on a vague but ultimately very important subject in physics, in which it is important if a particular phenomenon is *compatible* or *built-in*; if a phenomenon is *built-in* a theory, then the converse of this phenomenon is incompatible with the theory. One can say that a phenomenon is "accidental" in a theory if it is simply something that is *compatible* in the theory.

These concepts are best indicated with some examples. Newtonian physics has *built-in* the theory that all observers feel and observe the same passage of time: time is a universal parameter along a particle's worldline, a parameter that ticks on independent of your frame of reference. Therefore, Newtonian physics is *incompatible* with different observers feeling different passages of time. Newtonian physics also has *built-in* that all velocities are relative; Newtonian physics is *incompatible* with the existence of a universal speed that is the same for all inertial reference frames—the existence of the universal speed of light thus invalidates the fundaments of Newtonian mechanics.

Newtonian mechanics is *compatible* with the equality of m_{grav} and m_{inert}: we can simply set these quantities equal. General relativity, as we will see, has it *built-in* that these masses are equal. This also implies that general relativity is *not compatible* with the existence of a particle that has $m_{grav} \neq m_{inert}$—if such a particle should exist, then the fundaments of general relativity *must* be wrong.

This discussion still does not answer the question why it is preferable to have a theory where $m_{grav} = m_{inert}$ is *built-in* (like general relativity), rather than a theory that is simply *compatible* with this (like Newtonian physics).There is no general rule that tells us which physical phenomenon should be *built-in* the theory. Indeed, the equality of m_{grav} and m_{inert} was known for centuries (since Newton), without anybody really worrying about the fact that Newtonian mechanics did not explain this equality. It was Einstein who realized that $m_{grav} = m_{inert}$ is actually a profound, important statement that hints towards the special nature of gravity, and that a natural (and correct) theory of gravity would need to have this equality *built-in* to it.

If this discussion about *compatible* or *built-in* phenomena seems vague—it is. These concepts are ultimately related to the *feelings* that physicists have about theories and the "beauty" of a theory. A theory that explains a certain universal physical phenomenon (such as $m_{grav} = m_{inert}$) by having it *built-in* the theory is more *beautiful* than a theory that is simply *compatible* with this phenomenon. In a sense, the *built-in* theory *explains* the universality of the phenomenon. Another way to say this is that it is more beautiful to have a theory where "no accidents happen", in the sense that there are no accidental phenomenon that are *compatible* with the theory, but not *built-in* and thus not "explained" by the theory.

4.1.2 Other Principles of General Relativity

In the previous section, we have emphasized the (Einstein) *equivalence principle*:

Einstein's equivalence principle: All inertial observers are equivalent, where "inertial observer" means there are no forces working on the observer besides the gravitational force.

This is the most important principle that lies at the foundation of general relativity. This principle also replaces and subsumes the principle of relativity that we introduced earlier in special relativity, that said that all inertial observers (not including gravity) are equivalent.

Discussion 4.C: Mach's Principle

Consider the following questions:

(a) There is only one particle in the universe. Which of these concepts (as properties of the particle) make any sense: m, v, a, F?
(b) Try to define "standing still" and "moving at a constant velocity" very carefully.

There are a few other principles that inspired Einstein to some extent. One is called *Mach's principle*, which states:

Mach's principle: Inertia is the result of interactions between bodies.

This is a little vague. A precise definition of Mach's principle does not really exist, and one can encounter subtly different definitions depending on the source. For us, Mach's principle tells us that *it doesn't make sense to talk about acceleration if there is only one body (or particle) in the universe.* Think about it: if there is only one particle, is it meaningful to say it is accelerating? Accelerating with respect to what? Inertia and acceleration can only be defined *relatively*. Mach's principle (a slightly different version of it) is also sometimes used to explain why the stars look "fixed" in the sky when we "stand still" (as opposed to when we would whirl around, because then the stars would also whirl around): in essence, it is simply that we *define* our notion of "standing still" as being the frame in which the stars are "fixed"—there is no other way to define "standing still" other than in relation to other things.

Another principle that we want to mention is the *principle of covariance*, also sometimes called the principle of *general* covariance or principle of diffeomorphism covariance, which we will state as:

Principle of covariance: The *form* of physical laws are invariant under arbitrary coordinate transformations.

In other words, for any physical law, there should exist a mathematical formulation that *looks the same* in any coordinate system. What we mean by "looks the same" is that we should be able to write down the mathematical formulation of a physical law in such a way that we don't need to worry about which coordinate system we are using.

Exercise 4.3** Does the way we are used to writing down physical laws satisfy the principle of covariance? Think about an inertial particle in flat space travelling along a straight line $x^\mu(\tau)$ where τ is its proper time; we saw that it satisfies:

$$\frac{d}{d\tau} u^\mu = 0. \tag{4.9}$$

This is a mathematical formulation of (the special relativistic version of) Newton's first law or inertia principle. Is this formulation valid in any coordinate system? What about polar (or spherical) coordinates? We will revisit exactly this problem in Sect. 4.2.2.

Note that the principle of covariance is in some sense a more precise, mathematical way of implementing the Einstein equivalence principle; the two are definitely intertwined and should not be considered completely separate principles.

A final principle that we want to mention is the *correspondence principle*, which tells us:

> *Correspondence principle:* In the absence of gravity, physics should reduce to what we already know is correct (i.e. special relativity).

This seems almost like an obvious demand on any theory: when you get rid of the "new" stuff (in this case, gravity), the result had better be consistent with what you already know is true!

4.2 Mathematics of GR I: Covariant Derivatives

4.2.1 Introduction: Curved Spacetimes

Discussion 4.D: Curved Versus Flat Spaces

Consider a flat sheet of paper and a (beach) ball, which we take as an example of a curved space.

(a) Imagine you are an ant on the (flat) sheet of paper. (Let's also assume you are very nearsighted, so you're only really aware of what's going on immediately around you.) Do you know the paper is flat? What happens when you bend the paper? How can you, as the ant, tell if you are on the flat sheet of paper or actually on the ball?

(b) Consider a pair of ants that are either on the sheet of paper or the ball. They are standing a short distance away from each other on the surface. Then, they start walking along what they think are parallel straight lines. What would happen with the ants if they are on the sheet of paper, and on the ball? What happens with the distance between the ants?

(c) Draw a triangle on the paper and the ball. Is there a difference between the two triangles? Try measuring the angles in the triangle's corners.

Up until now, we have mainly worked in flat space(time), with metric given by:

$$ds^2 = -c^2 dt^2 + dx^2 + dy^2 + dz^2. \tag{4.10}$$

The most crazy thing we have done with this metric is introduce curvilinear coordinates (spherical or polar, depending on the spatial dimension):

$$ds^2 = -c^2 dt^2 + dr^2 + r^2 d\theta^2 + r^2 \sin^2\theta d\phi^2. \tag{4.11}$$

Although this metric looks a bit different in these coordinates, we are still familiar enough with these coordinates that we are confident they are just another way of describing flat spacetime.

In general relativity, we will no longer only have flat spacetimes. Curvature of spacetime due to matter will be the crucial ingredient that general relativity needs to determine how different objects influence each other gravitationally. For example, the Schwarzschild metric is probably the simplest metric in general relativity that we can write down that is curved:

$$ds^2 = -\left(1 - \frac{2M}{r}\right)c^2 dt^2 + \frac{dr^2}{1 - \frac{2M}{r}} + r^2 d\theta^2 + r^2 \sin^2\theta d\phi^2. \qquad (4.12)$$

This metric describes the spacetime curvature due to the presence of a mass at the origin $r = 0$. (See further on Sect. 4.6.1 about the Schwarzschild solution.)

Why is curvature of spacetime important for physics? In flat spacetime, an inertial particle travels on a *straight line* in spacetime. A particularly important property of straight lines in flat spacetime is that *two parallel straight lines never intersect*. This means that two inertial particles travelling parallel next to each other will never collide or drift farther apart. *The same is not true in curved spacetime!* When spacetime has curvature, straight lines *do* intersect (or drift farther apart). This makes sense when we talk about gravity: two particles falling towards the Earth will start to come closer and closer together, and will eventually intersect at the center of the Earth (assuming they can travel through the Earth's crust).

Clearly, the Schwarzschild metric (4.12) "looks different" than the flat spacetime metric (4.11), even though the coordinates used (t, r, θ, ϕ) are the same. However, the fact that a metric looks different than what we are used to is no guarantee that the metric is curved. For example, consider the metric (with coordinates t, x, y, z):

$$ds^2 = -(x^2 - y^2)\cos 2t \, dt^2 + 2\sin 2t \, (dy - x \, dt)(dx + y \, dt)$$
$$+ 2\cos 2t \, dt \, (x \, dy + y \, dx) + \cos 2t \, (dx^2 - dy^2) + dy^2 + dz^2. \quad (4.13)$$

This is not a diagonal metric as there are also terms that mix e.g. dt and dx; the metric matrix $g_{\mu\nu}$ is:

$$g_{\mu\nu} = \begin{pmatrix} -(x^2 - y^2)\cos 2t - 2xy\sin 2t & y\cos 2t - x\sin 2t & x\cos 2t + y\sin 2t & 0 \\ y\cos 2t - x\sin 2t & \cos 2t & \sin 2t & 0 \\ x\cos 2t + y\sin 2t & \sin 2t & 2\sin^2 t & 0 \\ 0 & 0 & 0 & 1 \end{pmatrix}. $$
$$(4.14)$$

Certainly, this metric looks very different than flat spacetime (4.10); however, it may come as a surprise that this metric is *flat*! It is simply *flat spacetime* in a contrived coordinate system!

> **Exercise 4.4***   Start with flat spacetime in Cartesian coordinates (4.10). Perform the coordinate transformation:
>
> $$x = x' \cos t' + y' \sin t', \quad ct = -x' \sin t' + y' \cos t', \quad y = y', \quad z = z',$$
> $$(4.15)$$
>
> and see that you get (after dropping the primes on the new coordinates) the metric (4.13).

If we know the coordinate transformation to bring a given metric to flat space in coordinates we are familiar with, as we did for (4.13), then we can safely conclude the metric is, indeed, flat. But what if we can't simply guess a good coordinate transformation? Is there some way of determining if a metric is curved or not, without having to resort to (guessing) coordinate transformations? Moreover, can we *quantify* curvature in some way; i.e. can we quantify if (a part of) a spacetime is *more or less curved* than another?

The answers to these questions lie in the mathematics of *differential geometry*. This is a branch of mathematics that deals with describing spacetimes (in mathematics, we call these objects *manifolds*) and distilling the curvature properties from the metric in certain ways. Differential geometry is then precisely the mathematical toolbox we need to formulate general relativity.

We will start our investigation of curved spacetimes and differential geometry by considering what seems like a deceptively simple problem: what is a straight line? Physically, a straight line is of utmost importance—in flat space we know that an inertial particle travels on a straight line. How do we really describe a straight line, mathematically? The derivative along a straight line is pretty simple (it is zero or constant, depending on how you think of the line). We will see that there is a natural way of generalizing this notion of constant/zero derivative, and of a "straight line", on curved manifolds and general metrics—mathematically, this requires the introduction of the *connection* and *covariant derivative*, as we will see.

4.2.2 Inertial Particle in Flat Space

Let's take a closer look at describing a very simple system: the motion of an inertial particle in flat $(2 + 1)$D spacetime (two spatial dimensions and one time dimension). Such a particle should simply travel on a straight line—so all we are doing is studying straight lines!

Cartesian and Polar Coordinates: An Apparent Contradiction?

Let's start by describing our particle in Cartesian coordinates $x^\mu = (ct, x, y)$, with metric:

$$ds^2_{cart} = -c^2 dt^2 + dx^2 + dy^2. \tag{4.16}$$

Say our inertial particle is travelling along a straight line at a constant velocity u in the x-direction. In Cartesian coordinates, this means:

$$x^\mu(\tau) = (c\gamma\tau, \ \gamma u\tau, \ y_0), \qquad \gamma = \frac{1}{\sqrt{1-(u/c)^2}}. \tag{4.17}$$

The particle's proper velocity is:

$$u^\mu = \frac{d}{d\tau}x^\mu = (c\gamma, \gamma u, 0), \tag{4.18}$$

so that $u^\mu u_\mu = -c^2$, as it should be for a massive particle. Since it is an inertial particle, the force working on the particle should be zero, which translates into the condition:

$$\frac{d}{d\tau}u^\mu = \left(\frac{\partial}{\partial x^\nu}u^\mu\right)\left(\frac{\partial x^\nu}{\partial\tau}\right) = (\partial_\nu u^\mu)\left(\frac{\partial x^\nu}{\partial\tau}\right) = u^\nu\partial_\nu u^\mu = 0, \tag{4.19}$$

which is of course satisfied.

In polar coordinates $\tilde{x}^\mu = (ct, r, \theta)$, with metric:

$$ds_{pol}^2 = -c^2dt^2 + dr^2 + r^2d\theta^2, \tag{4.20}$$

we now have (for the same particle moving in the x-direction):

$$\tilde{x}^\mu(\tau) = \left(c\gamma\tau, \ \sqrt{(\gamma u\tau)^2 + y_0^2}, \ \arctan\left(\frac{y_0}{\gamma u\tau}\right)\right). \tag{4.21}$$

Now, the proper velocity is:

$$\tilde{u}^\mu = \left(c\gamma, \ \frac{(\gamma u)^2\tau}{\sqrt{(\gamma u\tau)^2 + y_0^2}}, \ -\frac{y_0 u\gamma}{y_0^2 + u^2\gamma^2\tau^2}\right). \tag{4.22}$$

We still have $u^\mu u_\mu = -c^2$, as we should. However, notice that:

$$\frac{d}{d\tau}u^\mu = u^\nu\partial_\nu u^\mu \neq 0. \tag{4.23}$$

Does this mean there is a force acting on the particle? No—physics should not depend on the coordinates we use! We have simply described an inertial particle in flat space travelling along a straight line at constant velocity, in two different coordinate systems. The force the particle feels should not depend on the coordinate system we use!

From Cartesian to Polar Coordinates

Let's try to fix this by looking at the coordinate transformation from Cartesian to polar coordinates more carefully. The equation that the particle satisfies *in Cartesian coordinates* is:

$$\frac{d}{d\tau}u^\mu = \frac{d^2x^\mu}{d\tau^2} = 0.$$
(4.24)

Now, let's put in the coordinate transformation to polar coordinates (note that x^0 does not change):

$$x^0(\tau) = ct(\tau),$$
(4.25)

$$x^1(\tau) = x(\tau) = r(\tau)\cos\theta(\tau),$$
(4.26)

$$x^2(\tau) = y(\tau) = r(\tau)\sin\theta(\tau).$$
(4.27)

Let's plug in these expressions into (4.24). The x^0 equation is very simple:

$$\frac{d^2x^0}{d\tau^2} = c\frac{d^2t}{d\tau^2} = 0.$$
(4.28)

There's not much going on here. However, the spatial parts are more interesting; starting with $x^1 = x$:

$$\frac{dx^1}{d\tau} = \frac{d(r\cos\theta)}{d\tau} = \frac{dr}{d\tau}\cos\theta - r\sin\theta\frac{d\theta}{d\tau},$$
(4.29)

$$\frac{d^2x^1}{d\tau^2} = \frac{d}{d\tau}\left(\frac{dx^1}{d\tau}\right) = \cos\theta\frac{d^2r}{d\tau^2} - r\sin\theta\frac{d^2\theta}{d\tau^2} - 2\sin\theta\frac{dr}{d\tau}\frac{d\theta}{d\tau} - r\cos\theta\left(\frac{d\theta}{d\tau}\right)^2 = 0.$$
(4.30)

We have simply applied the chain rule many times to get these expressions. Analogously, we get for $x^2 = y$:

$$\frac{d^2x^2}{d\tau^2} = \frac{d^2y}{d\tau^2} = \sin\theta\frac{d^2r}{d\tau^2} + r\cos\theta\frac{d^2\theta}{d\tau^2} + 2\cos\theta\left(\frac{dr}{d\tau}\right)\left(\frac{d\theta}{d\tau}\right) - r\sin\theta\left(\frac{d\theta}{d\tau}\right)^2 = 0.$$
(4.31)

Now, we can take some particularly nice linear combinations of these equations and use trigonometry to get:

$$\cos\theta\frac{d^2x^1}{d\tau^2} + \sin\theta\frac{d^2x^2}{d\tau^2} = \frac{d^2r}{d\tau^2} - r\left(\frac{d\theta}{d\tau}\right)^2 = 0$$
(4.32)

$$-\sin\theta\frac{d^2x^1}{d\tau^2} + \cos\theta\frac{d^2x^2}{d\tau^2} = \frac{d^2\theta}{d\tau^2} + \frac{2}{r}\left(\frac{dr}{d\tau}\right)\left(\frac{d\theta}{d\tau}\right) = 0.$$
(4.33)

These equations are now equivalent to:

$$\frac{d}{d\tau}u^r - r\left(u^\theta\right)^2 = 0, \tag{4.34}$$

$$\frac{d}{d\tau}u^\theta + \frac{2}{r}u^r u^\theta = 0. \tag{4.35}$$

Now, we see where we went wrong! Our calculation shows us very clearly that, in polar coordinates, $du^\mu/d\tau$ does not vanish for a straight line; we must take into account the contribution of the second term in Eqs. (4.34)–(4.35).

Connection Coefficients

We can rewrite the Eqs. (4.34)–(4.35) together in this way:

$$\frac{d^2\tilde{x}^\mu}{d\tau^2} + \Gamma^\mu_{\nu\rho}\frac{d\tilde{x}^\nu}{d\tau}\frac{d\tilde{x}^\rho}{d\tau} = 0, \tag{4.36}$$

where all of the $\Gamma^\mu_{\nu\rho}$ are zero except:

$$\Gamma^r_{\theta\theta} = -r, \qquad \Gamma^\theta_{r\theta} = \Gamma^\theta_{\theta r} = \frac{1}{r}. \tag{4.37}$$

> **Exercise 4.5** Check that this is true, by explictly expanding out all of the summed over indices in polar coordinates!

The quantity $\Gamma^\mu_{\nu\rho}$ are called the *connection coefficients* or just *connection*; alternatively, it is also called the *Christoffel symbol* for the metric. It may seem like overkill to introduce these new symbols $\Gamma^\mu_{\nu\rho}$ (especially since most of the components are zero here, anyway). However, for other, more complicated coordinate transformations, we would need all of the connection coefficient to describe our equation of a straight line.

Using the basis vectors $\{\vec{e}_i\}$, there is a very simple way of understanding the Christoffel symbols:

$$\frac{\partial \vec{e}_\nu}{\partial x^\rho} = \Gamma^\mu_{\nu\rho}\vec{e}_\mu. \tag{4.38}$$

In other words, the Christoffel symbol $\Gamma^\mu_{\nu\rho}$ is the μ-component of $\partial\vec{e}_\nu/\partial x^\rho$; the Christoffel symbols describe how the basis vectors change as we move around.

Let's confirm this by working through the example for polar coordinates. The polar coordinate basis vectors are:

$$\vec{e}_r = \cos\theta\,\vec{e}_x + \sin\theta\,\vec{e}_y = \frac{\partial x}{\partial r}\vec{e}_x + \frac{\partial y}{\partial r}\vec{e}_y, \tag{4.39}$$

$$\vec{e}_\theta = -r\sin\theta\,\vec{e}_x + r\cos\theta\,\vec{e}_y. \tag{4.40}$$

A simple application of (4.38) for $\nu = \rho = r$ gives:

$$\frac{\partial \vec{e}_r}{\partial r} = 0 = \Gamma^{\mu}_{rr} \vec{e}_{\mu}, \tag{4.41}$$

which proves that $\Gamma^{r}_{rr} = \Gamma^{\theta}_{rr} = 0$. Next, set $\nu = r$ and $\rho = \theta$:

$$\frac{\partial \vec{e}_r}{\partial \theta} = -\sin \theta \, \vec{e}_x + \cos \theta \, \vec{e}_y = \frac{1}{r} \vec{e}_{\theta} = \Gamma^{\mu}_{r\theta} \vec{e}_{\mu}, \tag{4.42}$$

which gives us $\Gamma^{r}_{r\theta} = 0$ and $\Gamma^{\theta}_{r\theta} = 1/r$. Finally, for $\nu = \rho = \theta$, we have:

$$\frac{\partial \vec{e}_{\theta}}{\partial \theta} = -r \cos \theta \, \vec{e}_x - r \sin \theta \, \vec{e}_y = -r \, \vec{e}_r = \Gamma^{\mu}_{\theta\theta} \vec{e}_{\mu}, \tag{4.43}$$

which implies $\Gamma^{r}_{\theta\theta} = -r$ and $\Gamma^{\theta}_{\theta\theta} = 0$. We conclude that we have indeed reproduced (4.37) using (4.38)!

4.2.3 Connections and (Covariant) Derivatives

We have seen the transformation property for a coordinate transformation $x^{\mu} \rightarrow x'^{\mu}$ for a tensor with one downstairs (covariant) index:

$$K_{\mu} \rightarrow K'_{\mu} = \frac{\partial x^{\nu}}{\partial x'^{\mu}} K_{\nu}. \tag{4.44}$$

A good question to ask is: is the (partial) derivative of a function, $\partial_{\mu} f$, a tensor?

Exercise 4.6 Compare $\partial_{\mu} f$ for Cartesian coordinates and $\partial'_{\mu} f$ for polar coordinates, and see that these transform into each other in the correct way.

Another natural thing to do is to consider (partial) derivatives of other tensors, such as vectors v^{μ}. However, $\partial_{\mu} v^{\nu}$ is *not* a tensor!

Exercise 4.7 Prove that this is true: take a vector in $(2+1)$D Cartesian coordinates $v^{\nu} = (v^t, v^x, v^y)$, and its derivative $\partial_{\mu} v^{\nu}$; then prove that it is not a tensor by considering $\partial_{\mu} v^{\nu}$ in polar coordinates and the relation between the two quantities.

We would like to *improve* the derivative operator ∂_μ so that it always gives another tensor when applied to a tensor. We will call this improved derivative operator the *covariant derivative*, and denote it by ∇_μ (∇ is "nabla"). We found that $\nabla_\mu f = \partial_\mu f$ when acting on functions f, and we should have $\nabla_\mu = \partial_\mu$ in flat space in Cartesian coordinates.

Exercise 4.8* Why do we want $\nabla_\mu = \partial_\mu$ in flat space (in Cartesian coordinates)? Think of the principles of general relativity that we discussed above.

Exercise 4.9* If $\nabla_\mu V^\nu = \partial_\mu V^\nu$ in flat space in Cartesian coordinates, and $\nabla_\mu V^\nu$ transforms as a tensor, find $\nabla_\mu V^\nu$ in flat space in *polar* coordinates. See that you can express your answer as:

$$\nabla_\mu V^\nu = \partial_\mu V^\nu + \Gamma^\nu_{\mu\rho} V^\rho, \tag{4.45}$$

where $\Gamma^\nu_{\mu\rho}$ are precisely the coefficients we discussed in the previous section.

The *connection* or *connection coefficients* or *Christoffel symbols* $\Gamma^\mu_{\nu\rho}$ enters the covariant derivative of a vector as:

$$\nabla_\nu V^\mu = \partial_\nu V^\mu + \Gamma^\mu_{\nu\rho} V^\rho. \tag{4.46}$$

The connection coefficients are given by:

$$\Gamma^\mu_{\nu\rho} = \frac{1}{2} g^{\mu\sigma} \left(\partial_\nu g_{\rho\sigma} + \partial_\rho g_{\nu\sigma} - \partial_\sigma g_{\nu\rho} \right). \tag{4.47}$$

Properties of the Connection

Exercise 4.10 (*Basic properties*)

(a) Prove that the connection is symmetric in its lower indices, $\Gamma^\mu_{\nu\rho} = \Gamma^\mu_{\rho\nu}$.
(b) Use the expressions for the covariant derivatives of a scalar and a vector to derive an expression for $\nabla_\mu V_\nu$ (where V thus has a covariant index). Generalize to find an expression for the covariant derivative of a tensor with n contravariant and m covariant indices.
(c) Prove that the connection satisfies: $\nabla_\mu g_{\nu\rho} = 0$. A connection that satisfies this property is called a *Levi-Civita connection*. (This is the only type of connection we will use, but it is possible to define other connections.) Explain the consequence for equations like $\nabla_\mu V_\nu$ if V^μ is a vector.

Exercise 4.11* How does the connection transform? Either use the definition of $\Gamma^\mu_{\nu\rho}$ in terms of (derivatives of) the metric, or the fact that a covariant derivative of a vector transforms as a tensor, to find that the connection transforms under $x^\mu \to x'^\mu$ as:

$$\Gamma'^\mu_{\nu\rho} = \frac{\partial x'^\mu}{\partial x^\alpha} \frac{\partial x^\sigma}{\partial x'^\nu} \frac{\partial x^\lambda}{\partial x'^\rho} \Gamma^\alpha_{\sigma\lambda} + \frac{\partial x'^\mu}{\partial x^\sigma} \frac{\partial^2 x^\sigma}{\partial x'^\nu \partial x'^\rho}. \tag{4.48}$$

The first term in the transformation rule is what we would expect if $\Gamma^\mu_{\nu\rho}$ was a tensor; the second term involves second derivatives of the coordinate transformation *and tells us that the connection is not a tensor!*.

The result of the last exercise is important enough to repeat:

The connection $\Gamma^\mu_{\nu\rho}$ is *not a tensor*!

In fact, in a sense it is not surprising that the connection is not a tensor: we precisely introduced the connection to "improve" the partial derivative to be a tensor; therefore, it must compensate the "non-tensor" part of the partial derivative.

Using vector notation, we have, since $\vec{v} = v^\mu \vec{e}_\mu$ and using (4.38):

$$\frac{\partial \vec{v}}{\partial x^\nu} = \frac{\partial v^\mu}{\partial x^\nu} \vec{e}_\mu + v^\mu \Gamma^\rho_{\mu\nu} \vec{e}_\rho. \tag{4.49}$$

The covariant derivative then simply arises from factoring out $\vec{e}_\mu$ and relabelling dummy indices:

$$\frac{\partial \vec{v}}{\partial x^\nu} = \left(\frac{\partial v^\mu}{\partial x^\nu} + \Gamma^\mu_{\nu\rho} v^\rho \right) \vec{e}_\mu = \left(\nabla_\nu v^\mu \right) \vec{e}_\mu. \tag{4.50}$$

This is another way to see that the covariant derivative is the natural generalization of a derivative to a vector.

4.2.4 Parallel Transport

Discussion 4.E: Parallel Transport

Consider, once again, a flat sheet of paper and a beach ball. An ant at one point on each of these surfaces has a vector and wants to "drag it around" on the surface it walks on.

(a) If the ant drags around the vector on the flat sheet of paper, what seems to be the most natural way to have it dragged around "without changing" the vector? (Consider Fig. 4.1 if it is unclear what this question is referring to.)

(b) If the ant drags around the vector from any point A to any other point B on the flat sheet of paper in this natural way (i.e. "without changing" the vector), does it matter which path I take going from A to B?

(c) Now, consider a vector on the ball. Try to visualize or draw how a vector changes as the ant drags it along lines on the ball, again "without changing" the vector as the ant drags it along. (It's probably easiest to consider dragging the vector along great circles.)

(d) On the ball, does it matter which path the ant takes from any point A to any other point B? You can also try to have the ant drag the vector along a loop, i.e. a line that starts and ends at the same point. Does the vector always return to itself after going around the loop? (The easiest loop to consider is probably a triangle made out of sections of great circles on the ball.)

Let's say we have a vector with $V^y = 1$ (and other components zero) at the point $(1, 0)$ in the (x, y) plane. Now, we want to "move" this vector to the point $(2, 0)$ in the plane. There are obviously different ways we can do this (see Fig. 4.1), but the most natural way to do this seems to move it "without rotating" the vector. (We are using quotation marks to indicate that these concepts are not very rigorous.) If we move the vector in this way, it remains *parallel* to the original vector. We call this procedure *parallel transporting* a vector.

In this simple example, it was obvious what to do with the vector to parallel transport it. But for a general metric, and for curved manifolds, the concept of parallel transport might be more complicated; we will also see that this concept is intimately related to the concept of curvature on the manifold.

Let's try to formulate parallel transport in a way that we can generalize to general metrics and curved manifolds. We claim that:

$$\nabla_x V^\mu = 0, \tag{4.51}$$

is the equation that tells us that the vector V^μ is parallel transported along the x-direction.

Fig. 4.1 Moving the vector with $V^y = 1$ at the point $(1, 0)$ to the point $(2, 0)$ along a particular curve. There are different ways to do this, but the *parallel* one is the most natural

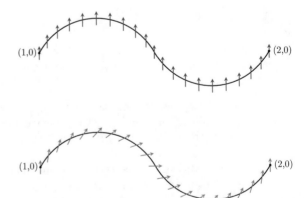

Exercise 4.12 How objects are parallel transported depends on the connection and thus depends on the metric. Different metrics will give different notions of parallel transport. Consider the vector ($V^r = 0$, $V^\theta = 1$) at the point ($r = 2$, $\theta = 0$) in (r, θ) coordinates on the plane defined in the regular fashion. For the regular polar coordinate metric:

$$ds^2 = dr^2 + r^2 d\theta^2, \tag{4.52}$$

try to see how this vector will be parallel transported around the circle $r = 2$. Now, consider the metric:

$$ds'^2 = dr^2 + d\theta^2. \tag{4.53}$$

For this metric in these coordinates, all Christoffel symbols $\Gamma^\mu_{\nu\rho} = 0$ (check this by calculating if you feel like it!). How will the vector be transported around the circle $r = 2$ now?

Exercise 4.13 For the Cartesian metric $ds^2 = dx^2 + dy^2$, convince yourself that (4.51) indeed determines the vector at $(x, y) = (2, 0)$ given the vector at $(1, 0)$, and that it indeed has been parallel transported to this point.

Exercise 4.14* Now do a coordinate transformation to polar coordinates $ds^2 = dr^2 + r^2 d\theta^2$. The expressions for the points $(x, y) = (1, 0)$ and $(x, y) = (2, 0)$ and the components of V^μ are more complicated now; but convince yourself that (4.51) indeed still transports the vector in a parallel fashion.

Exercise 4.15 Prove that if a vector V^μ is parallel transported along a given direction, then its magnitude $V^2 = V^\mu V_\mu$ is preserved along the transport.

Parallel transport becomes interesting and complicated when we consider it on a curved manifold. On a flat manifold, it is perhaps obvious that it does not matter how you parallel transport a vector from one point to another: the end result will be the same. A corollary is that if you parallel transport a vector away from a given point and then back to that point (i.e. parallel transport along a closed loop), then you will end up with the same vector again. The same is not true for curved manifolds: on a curved manifold, the *path of parallel transport* is important and different paths may have different resulting vectors; a corollary is that a vector parallel transported along a closed loop may not return to itself. An example of this on a sphere is shown in Fig. 4.2. We will see that this is precisely a measure of the curvature of a manifold.

Fig. 4.2 A vector is parallel tranported in a closed loop on the sphere; it does not return to itself

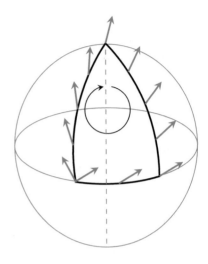

4.2.5   Geodesics

Discussion 4.F: Geodesics

We again consider our ant on a flat sheet of paper. The ant is an inertial observer.

(a) What kind of path does the ant take? What is the ant's velocity vector along the path?

(b) Consider the ant at a certain point along its path. Given its velocity vector on this point, do you have enough information to determine the next point in the ant's path? What determines the next point along the ant's path? Be as precise as you can! Is this true for any point along the ant's path?

(c) How does the ant's velocity change as it travels along its path? Use the language of parallel transport!

So far, this may have seen like a fairly trivial or silly exercise, since we know very well how an inertial observer in flat space behaves. However, being precise enough about the motion of the ant on a flat space will allow us to now fairly easily understand motion on *curved* space using the same concepts. So now let's consider an ant, who is an inertial observer on the sphere (a ball).

(d) What do we mean precisely by the ant being an "inertial observer"? Specifically, how should the ant's velocity change as it moves along its path?

(e) The ant is at a given point along its path. What determines the next point along the ant's path?

(f) Draw some possible paths for the ant on the ball. What kind of lines are these on the sphere?

> The ant's path is called a *geodesic*, which generalizes the concept of a *straight line* to curved spaces. Using the above ant thought experiment, summarize the key properties of a geodesic. (Hint: Make sure the properties tell you what determines the velocity or tangent vector at each point, and what determines how the curve itself looks.)

In the analysis we have done so far, we have essentially gathered all the intuition we need to answer one of the questions we started out with: how do we generalize the notion of a *straight line* in curved spacetimes and general metrics? In special relativity, an inertial particle travels on a straight line; we saw that the natural generalization of the equation defining a straight line was the equation:

$$\frac{d^2 x^\mu}{d\tau^2} + \Gamma^\mu_{\nu\rho} \frac{dx^\nu}{d\tau} \frac{dx^\rho}{d\tau} = 0, \tag{4.54}$$

where τ is the parameter along the path $x^\mu(\tau)$ (and $(dx^\mu/d\tau)(dx_\mu/d\tau) = -1$ for a timelike observer in spacetime). This equation also defines a straight line in flat space in other coordinates than Cartesian coordinates (such as spherical coordinates).

Equation (4.54) can also be used in more general, possibly curved, metrics. A path $x^\mu(\tau)$ that satisfies (4.54) is called a *geodesic*; a geodesic is the curved-space generalization of a straight line.

In flat space, a straight line is the line with the shortest distance between two points. A geodesic can be thought of *almost* generalizing this notion. On curved spaces, for two "sufficiently close points", a geodesic extremizes (so either minimizes *or maximizes!*) the proper distance between two points; we say a geodesic *locally* extremizes proper distance. A good example of this are geodesics on the (curved) sphere: these are (sections of) great circles, as depicted in Fig. 4.3. These extremize the distance travelled on a circle between two points as long as we don't go "too far around", because then it would be less far to go around the other way!

Fig. 4.3 A sphere with some geodesics drawn, these are (sections of) great circles on the sphere. If we go too far around the sphere along the geodesic, then the geodesic going around the other side can become of shorter distance

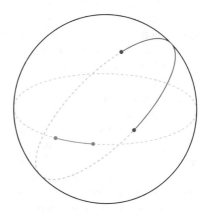

The geodesic equation (4.54) can also be written using the proper velocity vector $u^\mu = dx^\mu/d\tau$ (with $u^2 = -1$ for timelike geodesics):

$$u^\nu \nabla_\nu u^\mu = 0. \tag{4.55}$$

Exercise 4.16* Let's prove the equivalence of (4.55) and (4.54).

(a) First, use the chain rule to understand why:

$$\frac{d}{d\tau} f = u^\mu \partial_\mu f, \tag{4.56}$$

where f may have indices (that are not written). Hint: what is f a function of?

(b) Now write $d^2 x^\mu/d\tau^2 = d/d\tau(dx^\mu/d\tau)$ and prove equivalence of (4.55) and (4.54).

4.3 Principles of General Relativity, Revisited

The *principle of covariance* told us that the *form* of physical laws should be the same under coordinate transformations. Inspired by the mathematics we just learned, we can say that:

> All physical laws should be formulated using *tensors*. In particular, any *derivatives* appearing should be *covariant derivatives* ∇_μ.

The reason is that tensors (such as covariant derivatives) transform nicely under coordinate transformations; an equation that only involves tensors does not give different results for physics when we use different coordinate systems.

Speaking of covariant derivatives, the *correspondence principle* tells us that, when there is no matter and thus no gravity, spacetime must be flat, Minkowski spacetime since that is what special relativity tells us. It also implies that we must have $\nabla_\mu = \partial_\mu$ in flat spacetime, i.e. the correspondence principle picks out the connection given by (4.47), called the Levi-Civita connection, as the unique connection allowed in general relativity.

We can also consider Mach's principle. In essence, Mach's principle tells us that spacetime must be *flat, Minkowski spacetime* when there is no matter present. When there is no matter in spacetime, Mach's principle demands that there is no preferred direction or location in spacetime (think about this!); all points and directions have to be equivalent. This translates into a mathematical property of spacetime called being *maximally symmetric*, and it can be proven that the only spacetime that is maximally symmetric is flat, Minkowski spacetime. (Actually, there are a few other metrics that are also maximally symmetric; these are important in cosmology, see Sect. 4.6.4.) The converse of this thought also tells us that, if there *is* matter in spacetime, then spacetime should be *different* in different places and different directions. This strongly suggests that matter must *curve* spacetime; gravity *must* act by curving spacetime according to the matter present in it.

Einstein's equivalence principle told us that all inertial observers are equivalent, where "inertial" includes the possibility of a gravitational force. Mach's principle told us that the gravitational force acts by curving spacetime. Together, this means we can reformulate the equivalence principle more precisely:

> The laws of physics in *flat* and *curved* spacetime should be equivalent, i.e. their *form* should be the same.

For example, we saw that in flat spacetime, energy and momentum are conserved; we will see later that a similar conservation law must hold for curved spacetimes as well.

We know that, in flat space, inertial particles follow straight lines. We saw that a "straight line" in curved space is simply a geodesic that satisfies (4.54) or (4.55):

$$u^\nu \nabla_\nu u^\mu = 0. \tag{4.57}$$

Einstein's equivalence principle then immediately gives us:

> *The geodesic postulate:* An inertial (i.e. no forces acting on it except gravity, i.e. freely falling) particle will always travel on a *geodesic* in spacetime (satisfying (4.54) or equivalently (4.55)). A *massive* inertial particle will always travel along a *timelike* geodesic; this means $u^\mu = dx^\mu/d\tau$ always satisfies $u^2 = -1$. A *massless* inertial particle has $u^2 = 0$ and travels along a *null* geodesic.

The geodesic postulate is thus the natural (according to the equivalence principle) generalization of the principle of inertia in special relativity. As we have formulated it, this principle is indeed an *assumption* (i.e. a postulate) of the theory. However, an advanced treatment of general relativity can show that the geodesic postulate is actually a *consequence* of the rest of the theory; one can show that it follows from Einstein's equations—so strictly speaking, we do not actually need to assume it separately.

4.4 Mathematics of GR II: Curvature

We can now introduce the *Riemann tensor*, which will be the mathematical object that quantifies curvature.

4.4.1 Parallel Transport, Revisited

We saw in Fig. 4.2 that on a curved manifold, parallel transporting a vector in a closed loop (i.e. away from a point and then back to the point along a different path) does not necessarily result in the same vector we started out with. We can try to quantify this by considering the *commutator* of covariant derivatives along different directions. A commutator is the quantity:

$$[A, B] = A \cdot B - B \cdot A. \tag{4.58}$$

The most important property of a commutator is always:

$$[A, B] = -[B, A]. \tag{4.59}$$

We say that the commutator is *antisymmetric*. Let's consider the 2D plane and see what the commutator of two covariant derivatives gives, acting on a vector:

$$C = [\nabla_x, \nabla_y]V^\mu = \nabla_x \nabla_y V^\mu - \nabla_y \nabla_x V^\mu, \tag{4.60}$$

What does this quantity C teach us? We saw that a covariant derivative along a given direction tells us how something is parallel transported; thus, $\nabla_y \nabla_x V^\mu$ describes how the vector V^μ changes when we parallel transport it first (infinitesimally) along the x direction, and then transport it (infinitesimally) along the y direction. $\nabla_x \nabla_y V^\mu$ then describes how the vector V^μ changes when we parallel transport it first (infinitesimally) along the x direction, and then transport it (infinitesimally) along the y direction. The difference, $C = [\nabla_x, \nabla_y]V^\mu$, will then tell us the difference of the two vectors—if the manifold is flat, then it shouldn't matter which direction we do the parallel transport in first (see Fig. 4.4), and thus $C = 0$ on a flat manifold. Of course, we know that on the 2D flat plane with Cartesian coordinates, $\nabla_x = \partial_x$ and $\nabla_y = \partial_y$, so we indeed do have that:

$$C = (\partial_x \partial_y - \partial_y \partial_x)V^\mu = 0, \tag{4.61}$$

because partial derivatives commute.

Exercise 4.17** This example was fairly easy because all of the covariant derivatives ∇_μ were just partial derivatives ∂_μ in Cartesian coordinates. However, the plane should be flat in polar coordinates (r, θ) as well, with metric:

$$ds^2 = dr^2 + r^2 d\theta^2. \tag{4.62}$$

Calculate $C' = [\nabla_\mu, \nabla_\nu]V^\rho$ for some directions μ, ν in the plane, using polar coordinates. Now the covariant derivatives also have non-zero Christoffel symbols present, so the calculation is not so simple—however, convince yourself that $C' = 0$ no matter what directions μ, ν you chose.

Fig. 4.4 A vector is parallel transported from the origin in two ways to a given endpoint: first along the x direction and then y; or first along y and then x. Because the 2D plane is flat, the resulting vector is the same for both ways of parallel transportation

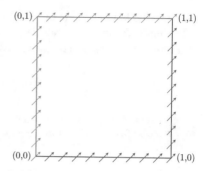

Now let's consider a curved 2D sphere with metric:

$$ds^2 = d\theta^2 + \sin^2\theta d\phi^2. \tag{4.63}$$

See Appendix A for the appropriate Christoffel symbols etc. Let's take the commutator $[\nabla_\theta, \nabla_\phi]$ acting on a vector. We can see that:

$$[\nabla_\theta, \nabla_\phi]V^\mu = \left(-\sin^2\theta V^\phi, V^\theta\right). \tag{4.64}$$

This is clearly not zero, which is consistent with our intuition that the sphere is curved. Take another look at Fig. 4.2.

We say that covariant derivatives or directions of parallel transport *don't commute* on a curved manifold. Equivalently, we can say that *parallel transport depends on the path taken* on a curved manifold.

Exercise 4.18 Prove (4.64) by explicit calculation.

4.4.2 The Riemann Tensor

We saw that the commutator of covariant derivatives tells us whether the manifold is curved or not. We can quantify the amount of curvature there is by the *Riemann tensor*, which is a tensor with four indices that can be defined as follows:

$$\nabla_\mu\nabla_\nu V^\rho - \nabla_\nu\nabla_\mu V^\rho = [\nabla_\mu, \nabla_\nu]V^\rho = R_{\mu\nu\sigma}{}^\rho V^\sigma. \tag{4.65}$$

Because the left hand side of (4.65) only contains partial derivatives and Christoffel symbols (in the covariant derivatives), the Riemann tensor can be completely determined in terms of only the Christoffel symbols and their derivatives:

$$R^{\mu}{}_{\nu\rho\sigma} = \partial_{\rho}\Gamma^{\mu}_{\nu\sigma} - \partial_{\sigma}\Gamma^{\mu}_{\nu\rho} + \Gamma^{\mu}_{\rho\lambda}\Gamma^{\lambda}_{\nu\sigma} - \Gamma^{\mu}_{\sigma\lambda}\Gamma^{\lambda}_{\nu\rho}. \tag{4.66}$$

Exercise 4.19* Prove that (4.66) satisfies (4.65).

As we have stressed before, the Riemann tensor is a measure for how flat a space(time) is. In fact, if we have $R_{\mu\nu\rho\sigma} = 0$, this implies that the space(time) is *precisely* equivalent to flat spacetime.

Exercise 4.20 Convince yourself that, even though $R_{\mu\nu\rho\sigma}$ is a tensor—which means that its components can change dramatically under coordinate transformations—that $R_{\mu\nu\rho\sigma} = 0$ is an equation that is unaltered by coordinate transformations. Therefore, if $R_{\mu\nu\rho\sigma} = 0$ in one coordinate system, there can be no coordinate transformations that makes the Riemann tensor non-zero.

4.4.3 Symmetries of the Riemann Tensor

The Riemann tensor, being a tensor with *four* indices, may seem like a scary beast of a tensor. The more indices, the more components a tensor has; we know that in four spacetime dimensions, a vector has four components; the metric (a symmetric matrix) has ten components—the Riemann tensor, having four indices, seems to have $4^4 = 256$ components! However, the Riemann tensor satisfies a number of symmetries that reduces the number of independent components—just like a symmetric matrix satisfies $M_{ab} = M_{ba}$, which reduces the number of independent components by half. The simple symmetry properties are:

$$R_{\nu\mu\rho\sigma} = -R_{\mu\nu\rho\sigma}, \qquad\qquad R_{\mu\nu\sigma\rho} = -R_{\mu\nu\rho\sigma}, \tag{4.67}$$
$$R_{\rho\sigma\mu\nu} = R_{\mu\nu\rho\sigma}. \tag{4.68}$$

The first line is called the *antisymmetry* of the Riemann tensor in interchanging either the first two or last two indices; the second line tells us the Riemann tensor is *symmetric* under interchanging the first *pair* of indices with the second pair.

> **Exercise 4.21***   Prove that (4.66) satisfies these symmetry properties. Also convince yourself that listing all three of the symmetry properties above is redundant; i.e. one of them follows already from the other two.

A more complicated property is called the *first Bianchi identity* for the Riemann tensor:

$$R_{\mu\nu\rho\sigma} + R_{\mu\sigma\nu\rho} + R_{\mu\rho\sigma\nu} = 0. \tag{4.69}$$

Using all of these symmetry properties, the amount of independent components of the Riemann tensor reduces to only 20.

> **Exercise 4.22****   Prove that the Riemann tensor with $4^4 = 256$ components only has 20 independent components due to the symmetry relations listed above. Hint: first consider the antisymmetry property (in both pairs of indices): how many relations is this? This should reduce the number of components to 32. Then take the symmetry property of interchanging index pairs—now there are only 21 left. Finally, argue that the first Bianchi identity only deletes one more independent component, giving a final total of 20.

One final differential equation that the Riemann tensor satisfies is called the *second Bianchi identity*:

$$\nabla_\lambda R_{\mu\nu\rho\sigma} + \nabla_\rho R_{\mu\nu\sigma\lambda} + \nabla_\sigma R_{\mu\nu\lambda\rho} = 0. \tag{4.70}$$

4.4.4 The Ricci Tensor and Scalar

The Riemann tensor, as we have tried to stress, is *the* object that quantifies how much (if any) a spacetime is curved. However, in general relativity, we often only use a tensor called the *Ricci tensor* $R_{\mu\nu}$, simply defined as follows:

$$R_{\mu\nu} = R^\rho{}_{\mu\rho\nu} = g^{\rho\sigma} R_{\rho\mu\sigma\nu}. \tag{4.71}$$

The Ricci tensor is thus defined by summing over one upstairs and one downstairs index of the Riemann tensor, or equivalently by contracting two downstairs indices with the metric. Note that we use the same letter R to denote the Riemann and Ricci tensors; by the number of indices present it should be clear which of the two tensors are being used.

The *Ricci scalar R* is defined by further contracting the Ricci tensor:

$$R = g^{\mu\nu} R_{\mu\nu} = R^{\mu}{}_{\mu}. \tag{4.72}$$

Once again, we use the same letter R for the Ricci scalar.

Finally, we can define the so-called *Einstein tensor* $G_{\mu\nu}$:

$$G_{\mu\nu} = R_{\mu\nu} - \frac{1}{2} g_{\mu\nu} R. \tag{4.73}$$

This particular combination of the Ricci tensor, metric, and Ricci scalar, is important for general relativity.

The most important property of the Ricci tensor is that it is symmetric, just like the metric is symmetric ($g_{\mu\nu} = g_{\nu\mu}$):

$$R_{\mu\nu} = R_{\nu\mu}. \tag{4.74}$$

An immediate consequence is that also the Einstein tensor is symmetric, $G_{\mu\nu} = G_{\nu\mu}$.

Exercise 4.23   Prove that the Ricci tensor is symmetric using the symmetry properties of the Riemann tensor. Also, using the symmetry properties of the Riemann tensor, convince yourself that the Ricci tensor is the only non-zero way of getting a tensor with two indices from the Riemann tensor (by contracting a metric $g^{\mu\nu}$ with the Riemann tensor $R_{\mu\nu\rho\sigma}$).

Exercise 4.24   If the Ricci tensor were antisymmetric, $R_{\mu\nu} = -R_{\nu\mu}$, would the above definition of the Ricci scalar make sense?

Exercise 4.25**   Using the second Bianchi identity for the Riemann tensor, prove that the Einstein tensor is *divergenceless*:

$$\nabla_\mu G^{\mu\nu} = 0. \tag{4.75}$$

4.4.5   Curvature and (Geodesic) Motion

In special relativity on flat spacetime, particles travel along straight lines. We saw that geodesics are the natural generalization of the concept of a "straight line" on curved spacetimes, and we discussed that general relativity postulates that inertial particles travel along these geodesics on curved spacetimes.

It is also interesting to consider *multiple* inertial particles in flat versus curved spacetime and their relative motion. In flat spacetime, two inertial particles simply

Fig. 4.5 Two geodesics (great circles) on the sphere; we see they start out parallel but end up converging in one point

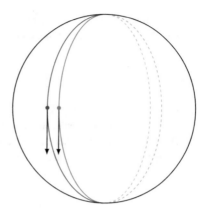

travel along two straight lines. If they are traveling parallel to one another at one time, they will always be travelling parallel to each other in the future. The distance between the two inertial observers will not change in time. This is no longer true on curved manifolds, as can be seen already with geodesics on the sphere, see Fig. 4.5.

Exercise 4.26* The concept of "parallel lines" in flat spacetime is not something that can be generalized or makes sense in curved spacetimes. Try to understand why; remember that a spacetime being curved is equivalent with *parallel transport* being dependent on the path that you transport over.

Since the way that two geodesics come together or drift apart is dependent on the curvature of spacetime, and the Riemann tensor carries all of the information about the curvature, we should be able to relate the drifting (or convergence) of geodesics to the Riemann tensor.

Exercise 4.27 Do two geodesics on the sphere (that start out on different points) drift apart or come together? We say the sphere has *positive* curvature. Can you think of what a spacetime would look like with *negative* curvature?

To quantify this relation between multiple geodesics and the Riemann tensor, we can consider a *family* of geodesics $x^\mu(\tau, \lambda)$. For every fixed λ (and τ varying), this gives one particular geodesic. When we change λ, we move along the family of geodesics to another one. For example, consider the geodesic at $\lambda = 0$ and $\lambda = 1$, which we can denote by $x_0^\mu(\tau) = x^\mu(\tau, \lambda = 0)$ and $x_1^\mu(\tau) = x^\mu(\tau, \lambda = 1)$; see Fig. 4.6.

The vector $u^\mu = dx^\mu/d\tau$ gives the proper velocity along a geodesic; this still remains true. Now, since we have another parameter λ, we can also define the vector:

$$n^\mu = \frac{dx^\mu(\tau, \lambda)}{d\lambda}. \tag{4.76}$$

The same way that u^μ describes the motion along the geodesic, n^μ describes how we move from one geodesic to another in the family of geodesics. Thus, the *change* of n^μ as we move along the geodesic family, i.e. the change of n^μ as we increase τ, will tell us something about how the geodesics in the family diverge or converge. The equation determining this is, in the notation of (4.55):

$$u^\nu \nabla_\nu \left(u^\rho \nabla_\rho n^\mu \right) = -R^\mu{}_{\nu\rho\sigma} u^\nu u^\rho n^\sigma. \tag{4.77}$$

Exercise 4.28**   Derive (4.77). You will need to use (4.54) to derive an equation determining $d^2 n^\mu/d\tau^2$. Remember also that (4.55) is equivalent to (4.54). You will also need (4.66) and (4.69) for the Riemann tensor.

So, the Riemann tensor precisely described how n^μ changes along a geodesic, which in turn describes how geodesics move closer or further away from each other.

Fig. 4.6 A family $x^\mu(\tau, \lambda)$ of geodesics, with two particular geodesics $x_0^\mu(\tau) = x^\mu(\tau, \lambda = 0)$ and $x_1^\mu(\tau) = x^\mu(\tau, \lambda = 1)$ indicated

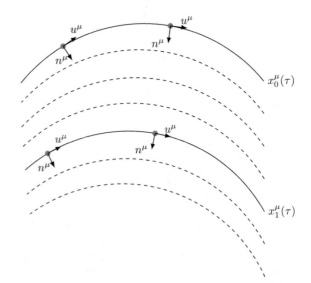

Exercise 4.29 Consider the sphere and the family of geodesics on it given by: $\theta(\tau, \lambda) = \tau$ and $\phi(\tau, \lambda) = \lambda$.

(a) Calculate u^μ and convince yourself that this is indeed a family of geodesics (with the parameter along the geodesics given by τ).
(b) Draw some of these geodesics on the sphere to get a feeling of what this family of geodesics looks like.
(c) Now consider n^μ and both sides of (4.77) (use the Riemann tensor expressions given in Appendix A); make sure they are equal! Try to understand this equation and how it relates to the evolution of the family of geodesics as τ increases.

Connections and Curvature in 1D and 2D

Exercise 4.30 (*One-dimensional accelerated frame*)
Consider a one-dimensional space with coordinate x. A straight line in this space, i.e. a trajectory of a particle travelling at constant velocity, is given by $x(t) = vt$. Now also consider a different coordinate system x', where $x' = \frac{a}{2}x^2$.
(a) For this straight line, what is $x'(t)$?
(b) Calculate $v^x = dx/dt$ and d^2x/dt^2, i.e. the velocity and acceleration in the x-coordinate system (with respect to the parameter t).
(c) Calculate $v^{x'} = dx'/dt$ and d^2x'/dt^2. Is this last quantity zero?
(d) Find the inverse relation giving x in function of x' (i.e. $x(x')$). Use this relation to relate dx to dx'. If the metric in the x-frame is $ds^2 = dx^2$, what is the metric in terms of dx'^2? What is the metric component g_{xx} and $g_{x'x'}$?
(e) Find the inverse metric $g^{x'x'}$ and calculate the (only possible component of the) connection using (4.47), i.e. $\Gamma^{x'}_{x'x'}$. You should find:

$$\Gamma^{x'}_{x'x'} = -\frac{1}{2x'}. \tag{4.78}$$

(f) Using the connection that you just calculated along with the values for dx'/dt and d^2x'/dt^2 that you found before, verify that (4.54) holds—i.e., $x'(t)$ is indeed a geodesic!
(g) Let's consider the basis vector $\vec{e}_{x'}$. This has as (only) component $e^{x'}_{x'} = 1$ (by definition). Using the metric ($g_{x'x'}$) that you found before, what is then $\vec{e}_{x'} \cdot \vec{e}_{x'} = |\vec{e}_{x'}|^2$? Use this to express $(\partial|\vec{e}_{x'}|)/(\partial x')$ in terms of $|\vec{e}_{x'}|$, and convince yourself that the result is consistent with the value for $\Gamma^{x'}_{x'x'}$ that you found before and (4.38).

Exercise 4.31 (*General one-dimensional coordinate*) Repeat the above exercise, but use a generic (inverse) coordinate relation $x = f(x')$ where f is an arbitrary function. Find the metric $ds^2 = dx^2$ in terms of dx' and find $\Gamma^{x'}_{x'x'}$. Write down the geodesic equation in the x' coordinate frame. You should find:

$$\frac{d^2 x'}{dt^2} + \partial_{x'}(\log f')\left(\frac{dx'}{dt}\right)^2 = 0. \tag{4.79}$$

Discuss why the Riemann tensor given in (4.66) is never important in a one-dimensional space. (Hint: what is the only possible value for all of the indices? What happens when you fill the indices in explicitly?)

Exercise 4.32 (*Two-dimensional spaces*) Consider the two (separate) metrics given by:

$$ds^2_{(1)} = e^{2x}dx^2 + e^{2x}dy^2, \qquad ds^2_{(2)} = \frac{1}{x^2}dx^2 + \frac{1}{x^2}dy^2. \tag{4.80}$$

Do the following for each of these metric separately:

(a) Find the components of the metric g_{xx}, g_{yy}. (Are there any other non-zero components?)
(b) Find the inverse metric, i.e. g^{xx}, g^{yy}.
(c) Use (4.47) to find the Christoffel symbols. You only need to calculate Γ^x_{xx}, Γ^y_{xy}, Γ^x_{yy} (and note that always $\Gamma^y_{yx} = \Gamma^y_{xy}$); all other components are zero (for both metrics).
(d) Write down the geodesic equation(s). How many equations do you need to write down?
(e) Use (4.66) to calculate $R^x_{\ yxy}$. Without calculating anything else explicitly, what are all other possible components of the Riemann tensor? What do you conclude about these two-dimensional spaces? Is one of them flat space?

4.5   The Einstein Equations

At this point, we have the necessary mathematical tools to describe curvature with the Riemann tensor (and its contractions, the Ricci tensor, Ricci scalar, and Einstein tensor). Einstein's equations are a mathematical statement that relates this curvature of spacetime to the matter in it. However, before being able to write down Einstein's equations, we need to first discuss how to capture the relevant properties of matter in the *energy-momentum tensor* $T_{\mu\nu}$.

4.5.1 The Energy-Momentum Tensor

We remember from special relativity that a particle's momentum four-vector p^μ describes the amount of energy and momentum the particle carries with it through spacetime. If we sum up all possible p^μ's of all possible particles, we get a four-vector p^μ that describes the *total* amount of energy and momentum flowing through any given point in spacetime.

The *energy-momentum tensor* $T^{\mu\nu}$ is a symmetric two-tensor (i.e. a symmetric matrix) that generalizes p^μ. In flat spacetime with normal Cartesian coordinates, the energy-momentum tensor is related to p^μ by:

$$T^{0\mu} = T^{\mu 0} \propto p^\mu . \tag{4.81}$$

That is, $T^{0\mu}$ will be zero at any point away from the particle, while at the position of the particle it is proportional to the four-momentum p^μ of the particle. Thus, $T^{00} \propto p^0$ is related to the energy density, and $T^{0i} = T^{i0} \propto p^i$ is related to the spatial component of the four-momentum.

The purely spatial components T^{ij} of the energy-momentum tensor give the *stress* or *pressure* ($T^{\mu\nu}$ is also sometimes called the *stress-energy tensor*). The diagonal components will be the most important to us: they describe the pressure in a given direction. Thus, if there is a pressure in the x direction, this is described by T^{xx}.

In general, we can say that $T^{\mu\nu}$ describes the flow of μ-momentum across a surface of constant x^ν.

Exercise 4.33* Check that this description indeed gives you $T^{0i} = T^{i0} \propto p^i$. Do you also understand why the diagonal components T^{ii} are the pressure? Can you describe and/or visualize the off-diagonal components, e.g. T^{xy}?

The most important property of the energy-momentum tensor is the *conservation of energy-momentum*:

$$\nabla_\mu T^{\mu\nu} = 0. \tag{4.82}$$

In special relativity, i.e. in flat spacetime, this equation simply becomes:

$$\partial_\mu T^{\mu\nu} = 0. \tag{4.83}$$

We can view this as the "master equation" for special relativistic kinematics; it contains e.g. the conservation of momentum for the $\nu = 0$ component:

$$\partial_\mu T^{\mu 0} = 0 \implies \partial_\mu p^\mu = \frac{1}{c^2}\frac{\partial E}{\partial t} + \left(\sum_i\right)\frac{\partial p^i}{\partial x^i} = 0. \tag{4.84}$$

The conservation equation (4.83) thus tells us that energy (and momentum) is conserved in flat spacetime, i.e. this is simply the principle of inertia. The curved spacetime equation (4.82) is the generalization of this principle of inertia; it tells us that energy and momentum is conserved in *curved* spacetimes as well.

We can be more precise about the energy-momentum tensor in the following way. If we have a particle with four-momentum p^μ located at position $\vec{x}_p$ in a flat spacetime, the energy-momentum tensor is given by

$$T^{0\mu} = T^{\mu 0} = cp^\mu \delta^{(3)}(\vec{x} - \vec{x}_p) , \qquad (4.85)$$

In this equation, the function $\delta^{(3)}$ is the three-dimensional *Dirac delta function*, which is defined such that $\delta^{(3)}(\vec{x} - \vec{x}_p)$ vanishes everywhere except at $\vec{x} = \vec{x}_p$, the position of the particle. This enforces our earlier condition $T^{0\mu}$ will be zero at any point away from the particle, while at the position of the particle it is proportional to the four-momentum p^μ of the particle.

Exercise 4.34   Start with the equation you get for conservation of energy, $dp^0/d\tau = 0$, in the reference frame of the particle. Convince yourself that in a general reference frame, (4.84) is the correct way (i.e. transforms as a scalar under Lorentz transformations) to generalize this equation for a general Lorentz frame.

Exercise 4.35   Say only the diagonal space-space components of $T^{\mu\nu}$ are non-zero. For example, T^{xx} is non-zero. Then (4.83) tells us that:

$$c\partial_0 p^x \delta^{(3)}(\vec{x} - \vec{x}_p) + \partial_x T^{xx} = 0. \qquad (4.86)$$

If T^{xx} is indeed the pressure in the x-direction, try to understand this equation from conservation of momentum (in the x-direction) and the fact that force is a derivative of pressure, i.e. $F^x = \partial_x p^x$.

Exercise 4.36 A *perfect fluid* or *ideal gas* is a fluid of non-interacting particles; in the rest frame of the fluid, this is completely described by its density and pressure at each point. The only non-zero components of $T^{\mu\nu}$ are $T^{00} = \rho c^2$ and $T^{xx} = T^{yy} = T^{zz} = p$.

(a) What conditions does (4.83) put on ρ, p?
(b) Convince yourself that the correct way to write the energy-momentum tensor in a way that is valid in any Lorentz frame is:

$$T^{\mu\nu} = \left(\rho + \frac{p}{c^2}\right) u^\mu u^\nu + pg^{\mu\nu}, \qquad (4.87)$$

where u^μ describes the four-velocity of the Lorentz frame we are using, and $g^{\mu\nu}$ is the inverse metric.

The main concept to remember is that the energy-momentum tensor $T^{\mu\nu}$ is a way to quantify the matter in the universe. Once we specify what the matter content of the universe is, there is a straightforward way to write down all components of the energy-momentum tensor. Each of these various components of this tensor should be thought of as encoding a particular piece of information (e.g. the energy density, momentum, etc.) about any particles, objects, or other things that are somewhere in our spacetime.

4.5.2 Einstein's Equations

Einstein's equations are essentially the summary of general relativity. It relates curvature (in the form of the Riemann tensor) to matter (in the form of the energy-momentum tensor), and is given by:

$$G_{\mu\nu} = R_{\mu\nu} - \frac{1}{2}g_{\mu\nu}R = \frac{8\pi G}{c^4}T_{\mu\nu}. \tag{4.88}$$

On the left hand (curvature) side of the equation, we have the Einstein tensor $G_{\mu\nu}$, which is a particular contraction of the Riemann tensor with metrics. On the right hand (matter) side, we have the energy-momentum tensor $T_{\mu\nu}$, and a constant of proportionality $8\pi G/c^4$.

Einstein's equation may seem arbitrary—why does this particular combination of the Riemann tensor appear on the left-hand side?—but is actually the unique equation one can write down, given a few basic assumptions. Those assumptions are:

1. The left hand (curvature) side of the equation should be *linear* in the Riemann tensor: that means there can be no terms like $R_{\mu\nu}R$ (which would be quadratic in the Riemann tensor).
2. Energy-momentum is conserved, i.e. $\nabla_\mu T^{\mu\nu} = 0$; this means the left-hand (curvature) side of the equation should satisfy $\nabla_\mu(\cdots) = 0$ as well.
3. For weak gravitational fields, (4.88) gives the same results as Newtonian gravity (which we know gives accurate results for weak gravitational fields). This fixes the constant of proportionality to be $8\pi G/c^4$, where G is Newton's gravitational constant (see later).

It turns out that the unique Riemann tensor combination that satisfies assumptions 1 and 2 is the Einstein tensor $G_{\mu\nu}$. However, as Einstein noticed, we have the freedom to add an extra term to the left-hand side proportional to the metric itself (i.e. strictly speaking, violating assumption 1), without violating assumption 2, which gives:

$$R_{\mu\nu} - \frac{1}{2}g_{\mu\nu}R + \Lambda g_{\mu\nu} = \frac{8\pi G}{c^4}T_{\mu\nu}. \tag{4.89}$$

Here, Λ is a *constant* called the *cosmological constant*. We are free to add this term because $\nabla_\mu g_{\nu\rho} = 0$, so energy-momentum is still conserved. Essentially, what adding the cosmological constant does is tell us that spacetime is curved *even when there is no matter (or energy) present*. Alternatively, we can move this term to the other side of the equation and view the cosmological constant as a kind of pervasive, omnipresent energy(-momentum)—in cosmology we speak of *dark energy*.

Exercise 4.37 We have introduced the Schwarzschild metric earlier (see also Appendix A). This metric is a solution to Einstein's equations (4.88). Given that $R_{\mu\nu} = 0$ everywhere for the Schwarzschild metric, what do you conclude about $T_{\mu\nu}$? Can you make sense of this? What about at $r = 0$?

4.5.3 Weak Field Approximation

We know that Newton's gravitational equation:

$$F^i_{grav} = \frac{d}{dt} p^i = -G \frac{m\tilde{m}}{r^2},\tag{4.90}$$

where m, $\tilde{m}$ are the masses of the objects and r is their separation, is a good description of a weak gravitational field. This means that Einstein's equations (4.88) had better be consistent with (4.90) when the gravitational field is weak; as we mentioned earlier, this fixes the constant of proportionality in Einstein's equations.

To show this, let us first consider a general constant of proportionality:

$$G_{\mu\nu} = \kappa \, T_{\mu\nu}.\tag{4.91}$$

We claim that the solution to this equation, given a point mass $\tilde{m}$ sitting at $r = 0$, is given by the Schwarzschild metric (4.12) with:

$$M = \frac{c^2}{8\pi} \kappa \, \tilde{m}.\tag{4.92}$$

(We will not verify this here; see later on in Sect. 4.6.1 and also Exercise 4.59.) Now, we assume the gravitational field is weak: this means that M/r is *small*, so that the metric is very close to Minkowski flat space-time. Let us now consider the motion of a particle with mass m in this spacetime. We know that its motion is governed by the geodesic equation:

$$\frac{d^2 x^\mu}{d\tau^2} + \Gamma^\mu_{\nu\rho} \frac{dx^\nu}{d\tau} \frac{dx^\rho}{d\tau} = 0.\tag{4.93}$$

We will also assume that the particle is travelling at very low speeds (compared to the speed of light)—this is necessary because Newton's gravitational law is also only valid for non-relativistic velocities. For a particle with $dx^\mu/d\tau = u^\mu = (c\gamma, \gamma v^i)$, this means that:

$$\frac{dt}{d\tau} \approx c, \qquad \frac{dx^i}{d\tau} \approx c\frac{dx^i}{dt} = cv^i. \tag{4.94}$$

We will disregard any terms that go as $O((M/r)^2)$ (because the gravitational field is weak) or $O((v^i/c)^2)$ (because the particle is travelling at low speeds). Then, the geodesic equation for the coordinate t is (see Appendix A for the Christoffel symbols of the Schwarzschild metric):

$$0 = \frac{d^2t}{d\tau^2} + \Gamma^t_{tr}\frac{dt}{d\tau}\frac{dr}{d\tau} \approx \frac{d^2t}{d\tau^2}, \tag{4.95}$$

because $\Gamma^t_{tr} \sim O((M/r)^2)$. This is, of course, consistent with the slow-moving approximation $dt/d\tau \approx c$. The geodesic equation for the coordinate r gives:

$$0 = \frac{d^2r}{d\tau^2} + \Gamma^r_{tt}\left(\frac{dt}{d\tau}\right)^2 + \Gamma^r_{ij}\frac{dx^i}{d\tau}\frac{dx^j}{d\tau} \approx \frac{d^2r}{dt^2}c^2 + \frac{M}{r^2}c^4. \tag{4.96}$$

So the equation we get, for a particle of mass m, is:

$$m\frac{d^2r}{dt^2} = \frac{d}{dt}(mv^r) = -c^2\frac{mM}{r^2} = -\kappa\frac{c^4}{8\pi}\frac{m\tilde{m}}{r^2}. \tag{4.97}$$

Comparing this to Newton's law (4.90), we conclude that we must have:

$$\kappa = \frac{8\pi G}{c^4}. \tag{4.98}$$

Exercise 4.38 There are a number of explicit calculations and approximations that are not spelled out in detail in the above derivation. Go through it and convince yourself everything works out correctly!

4.6 Solutions to Einstein's Equations

Now that we have the basic equations of general relativity, we can explore some of their solutions and see some of their interesting applications.

4.6.1   The Schwarzschild Solution

The Schwarzschild solution is probably the simplest possible solution (except flat space) to Einstein's equations. The Schwarzschild metric is given by:

$$ds^2 = -\left(1 - \frac{2M}{r}\right)(c^2 dt^2) + \frac{dr^2}{1 - \frac{2M}{r}} + r^2 d\theta^2 + r^2 \sin^2\theta d\phi^2. \qquad (4.99)$$

This is a solution to the *vacuum* Einstein equations, which means $T_{\mu\nu} = 0$ everywhere except at $r = 0$, where there is a mass m present. Note that the *geometric mass M* (which has units of length) is related to the actual mass m through:

$$M = \frac{G}{c^2} m. \qquad (4.100)$$

The quantity $2M$ is also called the *Schwarzschild radius*.

It can be proven that the Schwarzschild solution is actually the *unique* vacuum solution that is also *static* (i.e., it is both independent of time and looks the same if time were to flow backwards) and *spherically symmetric* (that is, there is no preferred direction). (See also Exercise 4.59.)

The Earth is not exactly spherically symmetric, and it certainly is not static (it rotates around its own axis, and also orbits the sun). However, for many purposes, the Schwarzschild metric with $m = m_{Earth}$ is a very good approximation of the spacetime curvature due to the Earth's mass at or above the Earth's crust.

Note that Appendix A contains a reference sheet of the Schwarzschild metric, including the explicit connection coefficients $\Gamma^{\mu}_{\nu\rho}$.

Exercise 4.39* We said that the Schwarzschild metric with $m = m_{Earth}$ is a good approximation for spacetime curvature *at or above* the Earth's crust. Why is the same metric *not* a good approximation for the spacetime curvature *inside* the Earth?

Exercise 4.40 (*Global Positioning System*) GPS works by communicating with satellites in orbit above the Earth. Essentially, a GPS user communicates with different satellites of which the position is known; the satellites then send a message telling the user the exact time on the satellite—by knowing the speed of light and measuring the difference in time stamps received between different satellites (and knowing precisely where the satellites are), the GPS user can triangulate their position very accurately. However, to be able to do this, one must be able to tell time aboard the GPS satellites with great accuracy, and in particular, one must take into account the relativistic time effects of the GPS satellites—both from special *and* general relativity! For the following, use that the mass of the Earth is about $m_E = 6.24 \cdot 10^{24}$ kg, and that its radius (to the

Earth's surface where we are) is about $r_E = 6520$ km. GPS satellites orbit the earth at an altitude (above the Earth's surface) of about $2 \cdot 10^4$ km.

(a) A GPS satellite travels in orbit around the Earth. We can model its orbital velocity with classical mechanics. To stay in orbit (and not fall to the Earth), the gravitational pull of the Earth (given by Newton's law) must be balanced by the centrifugal force of the satellite moving around the Earth, given by $m_{sat} v^2 / r$. Find the velocity of the satellite.

(b) Since the satellite is moving at such high speeds, we need to worry about relativistic time dilation! What is $\gamma - 1$ to leading order in u/c for the satellite?

(c) What is the time difference this implies between the Earth's surface and the satellite over a day, in seconds? Which clock ticks faster?

(d) Now, for the effect from *general* relativity! Use the Schwarzschild metric to calculate the difference of time elapsed per day on the surface of the Earth to time elapsed in the satellite's orbit. Which clock runs faster? *Hint*: Expand the time difference to first order in m_E.

(e) Are the special relativistic and general relativistic effects on the satellite's time both of the same sign (i.e. do they both slow down or speed up relative to the Earth)? Calculate the total time difference per Earth day between the satellite's time and the Earth's time due to both of these effects combined.

Exercise 4.41 The Earth's gravitational field (above the crust) is well approximated by the Schwarzschild metric. What is the Schwarzschild radius $2M$ for the Earth? (Remember, we found the Earth's mass and radius in Exercise 2.32.) Approximately how much have your feet aged more or less than your head in your lifetime?

Hint: Just plugging in numbers in a calculator can be a correct way to approach this problem, but some calculators may simply (incorrectly) output 0 due to numerical rounding errors. It can be helpful to first calculate the ratio squared of the rate of aging of your head divided by the rate of aging of your feet, and then use the Taylor expansion:

$$\frac{1}{r_0 + \epsilon} \approx \frac{1}{r_0} - \frac{1}{r_0^2}\epsilon, \tag{4.101}$$

followed by using the square root approximation (also a Taylor series):

$$\sqrt{1 + 2a\,\epsilon} \approx 1 + a\,\epsilon. \tag{4.102}$$

4.6.2   Black Holes

Remember that the Schwarzschild metric (4.99) describes the curvature of spacetime due to a mass centered at $r = 0$. When we get close enough to the mass, i.e. inside the Schwarzschild radius $r < 2M$, we enter a *black hole*—a place where even light cannot escape!

Black Holes Worksheet

Exercise 4.42 (*Black holes 101*) In this problem, we'll learn about some basics of black holes by looking at the Schwarzschild metric.

(a) Estimate the Schwarzschild radius $2M$ for a human and for the earth. Based on your answers, should we be worried about running into black holes in our everyday lives? (See also Exercise 4.39.)

(b) Suppose you shoot an object directly into a black hole such that the object's angular coordinates θ and ϕ are fixed and unchanging. Write down an expression for the infinitesimal length ds^2 between two points along the object's trajectory.

(c) Let's assume that the object you shot into the black hole is massive. When the object is outside the black hole, do you expect its trajectory to be timelike, spacelike, or lightlike? Why? What does this mean for the sign of ds^2? What does this mean for dr^2 after you cross the horizon at $r = 2M$? Can you turn around once you enter a black hole?

(d) Let's now suppose that you shoot a ray of light at the black hole, again with fixed angular coordinates θ and ϕ. What can you conclude about ds^2 along the trajectory of the light ray? Use this to to find dr/dt, the radial velocity of the light ray (as observed by you).

(e) What happens to the radial velocity of the light ray when $r \to \infty$? Does your answer make sense? What about when r approaches the Schwarzschild radius $2M$?

(f) Based on your answer to the previous question, what do you (standing far away from the black hole) see as the light ray approaches the horizon of the black hole?

Exercise 4.43 (*A Milky Way mystery*) Sagittarius A* is a massive object located the center of our Milky Way galaxy. For decades, astronomers have tried to figure out whether or not this object is a black hole by studying the orbits of stars around this object. In this problem, we will study the most modern data we have about Sagittarius A* and try to solve the problem ourselves!

(a) The star $S2$ orbits around Sagittarius A* in a smooth, elliptical orbit. The period T of this orbit is given by:

$$T = 2\pi\sqrt{\frac{a^3}{Gm}},\qquad\qquad(4.103)$$

where a is the semi-major axis of the orbit and m is the mass of the object $S2$ orbits around. Astronomers have measured the period and semi-major axis of $S2$'s orbit to be $T = 16$ years and $a = 1.5 \times 10^{14}$ meters, respectively. Estimate the mass of Sagittarius A*.

(b) Use your answer to part (a) to estimate the Schwarzschild radius $2M$ of Sagittarius A*.

(c) Out of all the stars that orbit Sagittarius A*, the one that comes the closest to it is $S14$. At the closest point in its orbit, it comes within 6.25 light-hours of Sagittarius A*. Convert this to meters to estimate the largest possible size that Sagittarius A* could have.

(d) The sun has a mass of $m_\odot = 2 \times 10^{30}$ kg. An object (or a collection of objects) is referred to as "supermassive" if it has a mass somewhere between 10^6 and 10^9 times that of the sun. Use your answer to part (a) to show that Sagittarius A* is a supermassive object.

(e) No supermassive stars have ever been discovered; any star with that much mass would simply collapse in on itself and form a black hole. Astronomers have therefore postulated that Sagittarius A* is either a supermassive black hole or a collection of many stars, each with a mass on the order of (or greater than) a solar mass (referred to as "main sequence" stars). In order for such a collection of stars to be stable, the total size of the collection would have to be significantly larger (i.e. tens of thousands or hundreds of thousands times larger) than its Schwarzschild radius. Is this true for Sagittarius A*?

(f) Based on your answers to part (e), argue whether or not Sagittarius A* is a black hole.

Exercise 4.44 (*Reissner-Nordström black holes*) We have so far we have looked at Schwarzschild black holes, which form when an uncharged (i.e. electrically neutral) star of mass m collapses in on itself. If the star that collapses is electrically charged, though, the result is a new type of black hole known as a Reissner-Nordström black hole. The metric is:

$$ds^2 = -\left(1 - \frac{2M}{r} + \frac{Q^2}{r^2}\right)(c^2 dt^2) + \frac{dr^2}{1 - \frac{2M}{r} + \frac{Q^2}{r^2}} + r^2 d\theta^2 + r^2 \sin^2\theta\, d\phi^2,$$

(4.104)

where M and Q both have units of length and are related to the mass m and charge q of the black hole by:

$$M = \frac{G}{c^2}m, \quad Q^2 = \frac{G}{4\pi\varepsilon_0 c^4}q^2.$$

(4.105)

(ε_0 is a constant that appears in electromagnetism. See Sect. 3.5.1 and e.g. (3.71).) The event horizon for these black holes is located at $r_{horizon} = M + \sqrt{M^2 - Q^2}$.

(a) Try to convince yourself that the horizon radius makes sense. There are many ways to argue this; one way would be to repeat Exercise 4.42, parts (c)–(f), but for the Reissner-Nordström metric (4.104).

(b) Suppose you know that a black hole has mass m. Given this, what range of values is the charge q of the black hole allowed to take? Justify your answer.

(c) An "extremal" Reissner-Nordström black hole is one with the maximum amount of charge q allowed for its mass m. Suppose you are outside of such an extremal black hole. In your hand, you hold an object whose mass m' is tiny and basically negligible, but whose charge q' is incredibly large. Is this object allowed to go into the black hole? What happens to the object if you throw it towards the black hole?

Exercise 4.45 (*One fish, two fish, redshift, blueshift*)   In this problem, we'll understand phenomena known as gravitational redshift and gravitational blueshift, where the presence of massive objects actually alters the properties of light waves nearby. You should first try to read Sect. 3.5.2 before attempting this problem; you will especially need (3.106). Understanding Exercise 3.70 first may help as well.

(a) Consider two observers, Alice and Bob, who are at distances r_1 and r_2 (respectively) away from the center of a Schwarzschild black hole of mass m. Assume that both Alice and Bob are stationary and do not change their position. Now, Alice uses a laser to shoot a photon to Bob. The photon initially has a frequency f_1 when Alice first sends it out. Prove that when Bob receives the photon, its frequency is now given by:

$$f_2 = f_1 \sqrt{\frac{r_2(r_1 - 2M)}{r_1(r_2 - 2M)}}. \tag{4.106}$$

Hint: if you're stuck, think back to how we calculated gravitational time dilation in Exercise 4.40. Derive a formula for how Alice's time related to Bob's time. Then, use this relation to determine how the photon frequencies are related.

(b) Suppose that Alice is further away from the black hole than Bob (i.e. $r_1 > r_2$). Does the photon's frequency increase or decrease as it travels to Bob? Use Fig. 3.12 to determine if the photon is "redshifted" or "blueshifted".

(c) Repeat part (b) for the case when Alice is closer to the black hole than Bob, $r_1 < r_2$.

(d) Suppose that Bob is hanging out right next to the black hole horizon. What happens to the frequency of the photon as it gets to Bob? What about the energy of the photon? Should you be worried about these results?

Let's investigate light-rays going into a black hole more carefully. First of all, note that $r = 2M$ is a special place for this metric: it does not seem like g_{tt} or g_{rr} really make sense here (they are either 0 or ∞, both of which are not allowed for a metric). However, this is what is called a *coordinate singularity*: something that is only singular (not well-behaved) in particular coordinates. We can do a coordinate transformation to get rid of a coordinate singularity; we introduce the tortoise coordinate $r^*(r)$:

$$r^* = r + 2M \ln \left| \frac{r}{2M} - 1 \right|, \qquad \frac{dr^*}{dr} = \left(1 - \frac{2M}{r} \right)^{-1}, \qquad (4.107)$$

and then we can use *ingoing Eddington-Finkelstein coordinates*:

$$v = ct + r^*, \qquad ds^2 = - \left(1 - \frac{2M}{r} \right) dv^2 + 2dv\, dr + r^2 (d\theta^2 + \sin^2 \theta d\phi^2). \tag{4.108}$$

In these coordinates, we can safely travel to $r = 2M$ and even $r < 2M$.

Now, we can see why light cannot escape from this black hole. Light travelling along the radial direction has $d\theta/d\lambda = d\phi/d\lambda = 0$, and so it has:

$$0 = - \left(1 - \frac{2M}{r} \right) \left(\frac{dv}{d\lambda} \right)^2 + 2 \frac{dv}{d\lambda} \frac{dr}{d\lambda}, \tag{4.109}$$

The solutions to this equation are:

$$\frac{dv}{d\lambda} = 0, \qquad \frac{dv}{d\lambda} = -\frac{dr}{d\lambda} r (2M - r). \tag{4.110}$$

The first class of solutions has $v = const.$, which means that as t increases, r^* (and thus r) *decreases*. These are *ingoing null geodesics* or *ingoing light rays*. The second class of solutions can be rewritten (using the chain rule and the above relation for dr^*/dr) to give:

$$\frac{dt}{d\lambda} = \frac{dr}{d\lambda} \left(\frac{r}{r - 2M} \right). \tag{4.111}$$

So, as $t(\lambda)$ increases, $r(\lambda)$ increases if $r > 2M$. These are *outgoing null geodesics* or *outgoing light rays*. However, we see that for $r < 2M$, $r(\lambda)$ must *decrease*; even the *outgoing* light rays are forced *inwards*—light cannot escape the black hole! The sphere at $r = 2M$ is called the *event horizon*: no communication from inside the event horizon to outside is possible. See Fig. 4.7 for the behavior of radial light rays in a black hole spacetime.

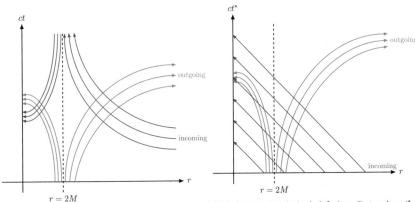

(a) Light ray trajectories in the (ct, r) coordinates of an observer standing somewhere (far) outside the black hole.

(b) Light ray trajectories in the (ct^*, r) coordinates where $ct^* = ct + 2M \ln \left| \frac{r}{2M} - 1 \right|$ so that ingoing light rays are straight lines at $45°$ angles. These are more natural coordinates e.g. for an observer falling into the black hole.

Fig. 4.7 Ingoing and outgoing light rays in the black hole spacetime, seen in two different coordinate systems. No light rays can exit the black hole

Exercise 4.46   Verify the coordinate transformations leading to the ingoing Eddington-Finkelstein coordinates.

Exercise 4.47**   There are also so-called *outgoing* Eddington-Finkelstein coordinates, given by:

$$u = ct - r^*, \qquad ds^2 = -\left(1 - \frac{2M}{r}\right) du^2 - 2du\, dr + r^2(d\theta^2 + \sin^2\theta d\phi^2).$$

$$(4.112)$$

Examine outgoing and ingoing null geodesics in these coordinates; what do you see? This is called a *white hole*.

Exercise 4.48*   A *curvature singularity* is a place where the curvature is so strong that general relativity predicts it is infinite: at such a point, general relativity does not really make any sense anymore! When do we have a *curvature singularity*, and when do we have a *coordinate* singularity? One test is to look at *curvature scalars*, which are scalars made from one or more Riemann tensors; if any such scalars are infinite at a point, then we are definitely dealing with a curvature singularity!

(a) Why is this the case? Hint: how does a scalar transform under coordinate transformations?
(b) The three simplest curvature scalars are R, $R_{\mu\nu} R^{\mu\nu}$, $R_{\mu\nu\rho\sigma} R^{\mu\nu\rho\sigma}$. Convince yourself these are the only possible curvature scalars you can construct using at most two Riemann tensors and the metric.
(c) What can you say about R and $R_{\mu\nu} R^{\mu\nu}$ everywhere for the Schwarzschild metric?

(d) For the Schwarzschild metric, we have:

$$R_{\mu\nu\rho\sigma}R^{\mu\nu\rho\sigma} = \frac{48M^2}{r^6}. \tag{4.113}$$

What can you conclude about the singularity at $r = 2M$? Where is there definitely a curvature singularity in the Schwarzschild metric?

Exercise 4.49** Try to understand what happens as you fall into a black hole. Here are some guiding questions:

(a) How will you *die*? Hint: Think of what happens if you are falling into the black hole feet-first; what part of your body is attracted more to the black hole? The difference in gravitational "force" between your head and feet is called *tidal forces*. You can quantify these using the methods in Sect. 4.4.5; where do these tidal forces become infinite (see also Exercise 4.48)?

(b) What will you *see* as you fall into a black hole? Try to understand (e.g. by drawing light rays) where the light comes from that you will see when you are outside the horizon, just at the horizon, and past the horizon.

(c) What will an external observer (sitting safely outside the horizon) see happen to you? Hint: Assume you are sending out light pulses as you fall into the black hole. What happens to the light pulses as you get nearer to and finally cross the horizon? (See also Exercise 4.42.)

(d) Once you cross the horizon at $r = 2M$, how long does it take you (from your own point of view) to fall and hit the curvature singularity (see Exercise 4.48) at $r = 0$? What about if you had a big engine strapped to you; could you slow your fall down and hit the singularity "later" if you pointed the engine away from $r = 0$ and turned it on? Hint: First, consider your proper velocity u^μ when you are falling towards the black hole radially; find a relation between u^r and u^t. Is $u^t = 0$ possible along the entire trajectory inside the horizon? Compare (qualitatively), for the situations $u^t = 0$ and $u^t \neq 0$, the time you feel elapse until you hit $r = 0$.

Penrose (Spacetime) Diagrams

We used spacetime diagrams in special relativity to graphically represent spacetime and indicate the relation between different reference frames. A similar but different concept in general relativity is that of a *Penrose diagram*. These are diagrams that depict spacetime as a *compact* (that is, with boundaries) object—a small diagram contains all the information about the entire *infinite* spacetime. An important property of Penrose diagrams is that (just as for the spacetime diagrams we used in special relativity) light rays always travel at 45° angles. However, it is usually not possible to draw clear axes indicating the coordinates in the diagram. We will not derive how to find these Penrose diagrams here, but we will give a few examples to show their usefulness.

Fig. 4.8 The Penrose
diagram for flat Minkowski
spacetime. See the text for
the description

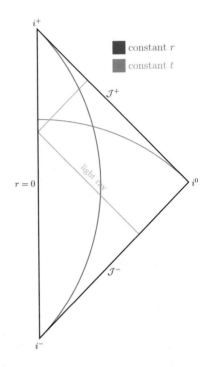

The Penrose diagram of flat, Minkowski spacetime is a simple triangle, see Fig. 4.8.
The rightmost point of the triangle is called *spatial infinity* (denoted by i^0); it is the
point that you get when $r \to \infty$ with all the other coordinates kept fixed. The top-
most (resp. bottom-most) point of the triangle is called *temporal infinity* (denoted
by i^+ resp. i^-), and that is reachable by taking $t \to \pm\infty$ while keeping all other
coordinates fixed. The other two slanted boundaries are "in between" these limits;
they are accessible by taking $r \to \infty$ and $t \to \pm\infty$ while keeping either $t + r$ or
$t - r$ fixed; these boundaries are called $\mathcal{J}^\pm$ (read $\mathcal{J}$ as "scri"). The vertical line on
the left is $r = 0$; anything that passes through here is "reflected". Remembering that
light always travels at 45° angles, we can have light originating at $\mathcal{J}^-$ and ending
on $\mathcal{J}^+$.

A (Schwarzschild) black hole has more features in its Penrose diagram, see
Fig. 4.9. First of all, notice that the structure "at infinity"—i.e. the partial triangle on
the right encompassing $i^\pm$, $\mathcal{J}^\pm$—is again present. This makes sense: far away from
the black hole—and especially "at infinity"—we expect there to be no effect left on
spacetime from the mass of the black hole. We say the metric is *asymptotically flat*.

Now, let's focus on region I and II in Fig. 4.9: Here, the horizon $r = 2M$ is given by
the upper-left slanted line. From the diagram (and keeping in mind that light travels
at 45° degree angles), it is immediately obvious that anything (including light) that
is inside the horizon, cannot "escape" to infinity. Rather, anything inside the horizon
must be carried to the squiggly line on top: the *singularity* at $r = 0$. This is a *curvature*

Fig. 4.9 The Penrose diagram for a Schwarzschild black hole. See the text for the description

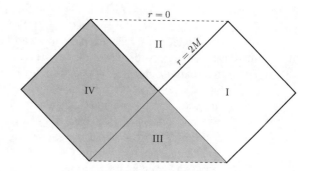

singularity (see also Exercise 4.48). General relativity breaks down here: we don't know anymore what happens when anything hits a curvature singularity! Thankfully, the singularity is "shielded" by the event horizon: standing outside the horizon, we can never see what happens at the singularity. This principle is also called the *cosmic censorship conjecture*: the conjecture that all such curvature singularities *must* be hidden behind an event horizon.

We will not go into much detail on regions III and IV in Fig. 4.9. A careful analysis of the Schwarzschild solution tells us that it actually also contains two regions which are an exact copy of the regions I and II, *except* that the (coordinate) time t "moves backwards". There is a wormhole that connects regions I and IV through the center of the diagram (which is called the "bifurcation sphere" at $r = 2M$), but this is a *non-traversable* wormhole: any line connecting points of region I and IV going through it will necessarily be *spacelike*.

4.6.3 Wormholes

Discussion 4.G: **Wormholes**

Consider the metric:

$$ds^2 = -c^2 dt^2 + dl^2 + (l^2 + n^2)(d\theta^2 + \sin^2 \theta d\phi^2). \qquad (4.114)$$

(a) First of all, what is this metric when $n = 0$? (Remind yourself what the metric of flat 3D space looks like in spherical coordinates, see (3.35) in Exercise 3.32). What is the allowed range of the coordinate l?
(b) Now consider $n \neq 0$. In the limit $l > 0$ for large l, what does the spacetime look like?
(c) Is $l < 0$ allowed when $n \neq 0$? What can you say about the spacetime for negative (and large) l?

(d) Now consider $l = 0$ (at an instant $t = t_0$ in time, to make things easier). Compare the situation for $n = 0$ and $n \neq 0$. What is $l = 0$ when $n = 0$? How many dimensions does this particular part of spacetime have? Now answer the same questions about the $l = 0$ part of spacetime when $n \neq 0$.

(e) Try to give a cartoon drawing of the spacetime (or more precisely, an instant in time of the spacetime) for $n \neq 0$, making a smooth drawing connecting what you know from the previous questions about the spacetime at $l = 0$, large $l > 0$, and large $l < 0$. You have (hopefully) just drawn a *wormhole*.

(f) Can you think of the wormhole we have been describing and drawing so far as an actual "shortcut" between two points in spacetime? Can you think of a way we might be able to remedy this to actually get a shortcut in spacetime?

(g) For this metric, we can calculate the Einstein tensor. In particular, we have:

$$G_{tt} = -\frac{n^2}{(n^2 + l^2)^2}. \tag{4.115}$$

Think about the Einstein equations. What does this value for G_{tt} mean for T_{tt}? What does T_{tt} represent physically? Do you think this value for T_{tt} makes any sense?

Wormholes are particular exotic solutions to Einstein's equations. We will focus on the so-called *Ellis wormhole* (see Fig. 4.10), with metric:

$$ds^2 = -c^2 dt^2 + dl^2 + (l^2 + n^2)(d\theta^2 + \sin^2\theta d\phi^2). \tag{4.116}$$

This looks just like the metric of flat space (with r replaced by l) when $n = 0$. When $n \neq 0$, the point $l = 0$ is no longer just a point (i.e. the origin)—it is now a sphere of radius n; we can use this sphere to travel *through* $l = 0$ and go to *negative* values of l. Also note that, for large (negative or positive!) values of l, we can neglect the n^2 term and we (approximately) have flat space. Thus, the Ellis wormhole glues two (approximately) flat spacetimes (those at positive and negative l) together through a *wormhole* at $l \approx 0$!

However, there is a big catch to this (and any other "traversable") wormhole! We can calculate the Einstein tensor $G_{\mu\nu} = R_{\mu\nu} - \frac{1}{2}g_{\mu\nu}R$, and the non-zero components are given by:

$$G_{tt} = G_{ll} = -\frac{n^2}{(n^2 + l^2)^2}, \qquad G_{\theta\theta} = \frac{G_{\phi\phi}}{\sin^2\theta} = \frac{n^2}{n^2 + l^2}. \tag{4.117}$$

Again, we can see that as l^2 gets big, these components get very small. However, for l small, we see that $G_{tt} < 0$, which implies through Einstein's equations that $T_{tt} < 0$. This means we have a *negative* energy density present! No matter or energy that we

Fig. 4.10 A graphical representation of the Ellis wormhole. The "throat" is when l is small; the two asymptotic regions are when l is large and negative *or* positive ("Ellis Wormhole Catenoid" by Turningwoodintomarble is licensed under CC BY 2.0)

Fig. 4.11 A wormhole acting as a shortcut between two areas in the same spacetime ("Wormhole Demo" by Panzi is licensed under CC BY 2.0)

know of so far in the universe has a negative energy density; that is why we think (traversable) wormholes probably don't exist.

The above wormhole might seem a little disappointing: after all, it connects two *different* (flat) spacetimes with each other; wasn't a wormhole supposed to be a kind of "shortcut" between two points in the *same* spacetime?! Well, we can easily imagine a situation where the two spacetimes connected by the (Ellis) wormhole are really one and the same; see Fig. 4.11. We simply need to alter spacetime so that it "bends" back towards itself!

The expression $T_{tt} < 0$, indicating a negative energy density, is not a very covariant one. A more covariant demand on spacetime is demanding that:

$$T_{\mu\nu}u^{\mu}u^{\nu} \geq 0, \tag{4.118}$$

for all non-spacelike (i.e. timelike or null) vector fields u^{μ}. This demand is exactly the demand that *there cannot exist any observer that measures a negative energy density in their reference frame*. This is called the *weak energy condition*.

Exercise 4.50 Prove that (4.118) indeed means that all observers measure a positive energy density.

Exercise 4.51* There are also other energy conditions other than the weak energy condition. The *strong energy condition* says that:

$$\left(T_{\mu\nu} - \frac{1}{2} T^{\rho}_{\ \rho}\, g_{\mu\nu} \right) u^{\mu} u^{\nu} \geq 0, \qquad\qquad (4.119)$$

for all non-spacelike vectors u^{μ}. The *dominant energy condition* says that:

$$- u^{\nu} T_{\nu}^{\ \mu} \qquad\qquad (4.120)$$

must be a future-directed non-spacelike vector for all future-directed timelike vectors u^{μ}.

(a) Prove that the dominant energy condition implies the weak one.
(b) Prove that the strong energy condition implies the weak one.
(c) Prove that the dominant energy condition does not imply the strong one, nor vice versa.
(d) Try to understand the physical significance of the dominant energy condition by understanding what the vector $-u^{\nu} T_{\nu}^{\ \mu}$ physically represents, for an observer with proper velocity u^{μ}.

The physically important energy condition is the dominant one. The physical meaning of the strong energy condition is not clear, but it is needed to prove important (singularity) theorems about black holes. A so-called inflationary universe (see below in Sect. 4.6.4) violates the strong energy condition. Note that in quantum field theory, all energy conditions are violated to some extent.

4.6.4 Cosmology

It is not hard to believe that the the study of the stars and the universe began with the ancients who gazed at the night sky wondering what the heavens contained. The natural curiosity of the human race has posed similar questions throughout the history of time. Questions like "Where did we come from?", "How did the planet we live on come to be?", and "How is the universe changing?" have been asked since humans have been capable of rational thought and the study of *cosmology* seeks to answer these questions. Cosmology is the study of the origin and development of the universe.

A few assumptions need to be made before one can start using math to describe the universe. The first of these assumptions is called the *cosmological principle*. This principle assumes that the universe is *homogeneous* and *isotropic* at large scales. Homogeneity means that there is no preferred location in the universe; on average, we expect to find approximately the same distribution of matter etc. at any location. Isotropy implies that there is no preferred direction in the universe; on average, we expect to see approximately the same in every direction we can observe. See Fig. 4.12 for graphical representations of homogeneity and isotropy.

The second assumption is a little more subtle; *Weyl's postulate* assumes that all particles that make up the substratum (the perfect fluid that is the universe) move on

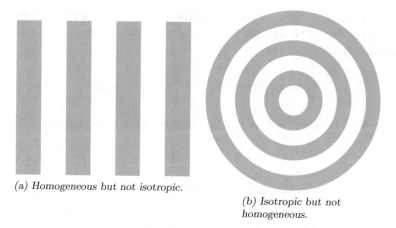

(a) Homogeneous but not isotropic.

(b) Isotropic but not homogeneous.

Fig. 4.12 Two patterns that demonstrate homogeneity and isotropy. The image on the left is homogeneous but not isotropic and the image on the right is isotropic but not homogeneous on large scales

time-like geodesics that diverged from a common point in the past (e.g. the Big Bang). In essence, Weyl's postulate means that there are (approximately) no intersections or interactions of e.g. galaxies other than in this common point in the past. Weyl's postulate tells us there are so-called "privileged observers" that are co-moving with the large-scale motion of (groups of) galaxies.

The last assumption we need to make is simply that the theory of general relativity works on the enormous large scales of the universe. Given these assumptions, we can attempt to describe the geometry of the universe in the context of general relativity.

Spaces of Constant Curvature and the Friedmann-Robertson-Walker Metric
To describe the evolution of the universe, we will first need to describe the geometric space that it will be modeled by. First of all, the geodesics in this space must be orthogonal to the spacelike hypersurfaces of the geometry. This means that the geodesics are orthogonal to the surfaces given by $t = const.$; this is a consequence of Weyl's postulate. Next, because we assume the universe is homogeneous and isotropic, the metric of the universe can only permit *constant curvature*. This means the Riemann tensor must take the form:

$$R_{\alpha\beta\mu\nu} = K(g_{\alpha\mu}g_{\beta\nu} - g_{\alpha\nu}g_{\beta\mu}), \qquad (4.121)$$

where K is a constant. Now, invoking homogeneity and isotropy, the metric must be *spherically symmetric*. This means that the metric has angular (θ, ϕ) components that look like a sphere; the term in the radial direction is still undetermined. Thus, the spatial metric must take the form:

$$ds^2 = e^\lambda dr^2 + r^2(d\theta^2 + \sin^2\theta d\phi^2), \qquad (4.122)$$

where $\lambda(r)$ is some a priori undetermined function of r. We can compare the Ricci tensor derived from (4.121) and the Ricci tensor derived from (4.122) to find the coupled equations:

$$\frac{\lambda'}{r} = 2Ke^\lambda, \qquad 1 + \frac{1}{2}re^{-\lambda}\lambda' - e^{-\lambda} = 2Kr^2, \qquad (4.123)$$

where we use the notation $\lambda' = d\lambda/dr$. These equations are solved by:

$$e^{-\lambda} = 1 - Kr^2. \qquad (4.124)$$

The full metric, including time, will then be:

$$ds^2 = -c^2dt^2 + R(t)^2 \left(\frac{dr^{*2}}{1 - kr^{*2}} + r^{*2}(d\theta^2 + \sin^2\theta d\phi^2) \right), \qquad (4.125)$$

where $R(t)$ is a function of time with units of length, r^* is the dimensionless variable

$$r^* = |K|^{\frac{1}{2}}r, \qquad (4.126)$$

and k can take the values $-1, 0,$ or 1, depending on if the universe is, respectively, *open, flat,* or *closed*. There are a few coordinate changes needed to get from (4.122) to (4.125), so k is a constant that is determined by but is not the same as K. The metric (4.125) is called the Friedmann-Robertson-Walker metric or FRW metric for short. With the inclusion of time, we also allow for the spatial part of the metric to depend on time through the function $R(t)$; this function is called the *scale factor*. The scale factor is not quite the naive radius of the universe, but does dictate the overall scaling of the universe.

The Friedmann Equation

The *Friedmann equations* are obtained by plugging the FRW metric into Einstein's equations with the particular energy-momentum tensor:

$$T_{\mu\nu} = \left(\rho + \frac{p}{c^2} \right) u_\mu u_\nu + pg_{\mu\nu}. \qquad (4.127)$$

This is the energy-momentum tensor for a perfect fluid (with density ρ and pressure p; see also Exercise 4.36); in the co-moving frame (using homogeneity and isotropy) $u^\mu = (c, 0, 0, 0)$. The Friedmann equations are then the two independent equations following from Einstein's equations (allowing the presence of a cosmological constant Λ):

$$\left(\frac{\dot{R}}{R}\right)^2 + \frac{kc^2}{R^2} = \frac{8\pi G\rho}{3} + \frac{\Lambda c^2}{3} \tag{4.128}$$

$$\frac{2\ddot{R}}{R} = -\left(\frac{\dot{R}}{R}\right)^2 - \frac{kc^2}{R^2} + \Lambda c^2 - \frac{8\pi Gp}{c^2}. \tag{4.129}$$

One semi-common convention is to combine these two equation into one *Friedmann equation*, which involves integration and the assumption that the universe is completely comprised of dust with zero pressure $p = 0$; this leads to the equation:

$$\dot{R}^2 = \frac{8\pi GR^2\rho}{3} + \frac{\Lambda c^2 R^2}{3} - kc^2. \tag{4.130}$$

The Friedmann equations are the basic tools necessary to make models of the universe; in fact, all possible outcomes for the fate of the universe are necessarily included in these equations.

We will discuss one example of a toy universe, a flat universe filled with dust and having no cosmological constant. Flatness means $k = 0$, while dust gives $p = 0$ and zero cosmological constant is $\Lambda = 0$. We will consider a matter-dominated universe (i.e. matter is the most important contribution to the total energy density), which can be shown to give $\rho = \rho_0 R^{-3}$. Then (4.130) becomes:

$$\dot{R}^2 = \frac{8\pi G\rho_0}{3R}, \tag{4.131}$$

which can be rewritten in the more convenient form:

$$R^{1/2}dR = \sqrt{\frac{8\pi G\rho_0}{3}}dt. \tag{4.132}$$

We can integrate both sides of (4.132) to find:

$$R(t) = \tilde{R}\,t^{2/3}. \tag{4.133}$$

This equation thus governs the expansion of a flat dust filled universe that is matter-dominated with no cosmological constant. We see that the universe expands with a fractional power of time. The scale factor of the universe will never be zero and it will continue to expand forever.

One can continue to find different models of the universe by solving the Friedmann equations given different parameters, although the integrals involved can get trickier. Depending on the model used, the universe could expand forever at a constant rate; it could expand up to a finite time and then begin to shrink and eventually collapse (in a *Big Crunch*); or, it could expand forever at an accelerating rate, eventually causing all matter to lose causal contact with each other (called the *Big Rip*). Modern cosmological experiments are aimed at measuring parameters of the universe to determine which of these fates will befall our universe.

There are many more topics to explore in cosmology. One such topic is the study of *dark matter* and *dark energy*, which combine to make up 95% of the universe but whose nature are as of yet completely unknown. Other notable topics are the study of when the first stars formed in the so-called *epoch of re-ionization*; the study of background radiation called the *cosmic microwave background* or CMB; the study of how large-scale structure in the universe formed; and the study of *inflation*, a period in the early universe where the universe is believed to have expanded by a factor of 10^{26} in about 6^{-35} s. There are many questions still to be answered about the universe we live in.

Exercise 4.52 Convince yourself that an isotropic and homogeneous universe requires a metric that admits a constant curvature. Qualitatively, what would happen to a uniform matter distribution if the curvature were to vary with position?

Exercise 4.53 Confirm that (4.121) above admits a constant curvature, i.e. show that the Ricci scalar is constant.

Exercise 4.54 Derive the FRW metric (4.125). Use the constant-curvature metric and add in time, including an unscaled scale factor, $S(t)$, to arrive at

$$ds^2 = -c^2 dt^2 + S(t)^2 \left(\frac{dr^2}{1 - Kr^2} + r^2(d\theta^2 + \sin^2\theta d\phi^2) \right). \qquad (4.134)$$

Now make the following transformation:

$$r^* = |K|^{\frac{1}{2}} r. \qquad (4.135)$$

The idea behind this transformation is to hide the magnitude of K in the radial coordinate and the scale factor. Let $K = |K|k$ and use the following definitions to arrive at the FRW metric:

$$R(t) = \begin{cases} S(t)/|K|^{\frac{1}{2}}, & (K \neq 0), \\ S(t), & (K = 0). \end{cases} \qquad (4.136)$$

Finally, using $K = |K|k$, convince yourself that the only allowed values of k are $-1, 0$, and 1.

Exercise 4.55 Consider a dust filled flat universe with a cosmological constant. How does the scale factor evolve with time? Assume $\rho = \rho_0$ at all times (this corresponds to a dark energy dominated universe, i.e. where the main contribution to the energy density is from the cosmological constant). How is this different from (4.133)?

4.6.5 Gravitational Waves

In general relativity, the spacetime distance ds between any two points that are infinitesimally close to one another is determined by the metric tensor $g_{\mu\nu}$:

$$ds^2 = g_{\mu\nu}dx^\mu dx^\nu . \tag{4.137}$$

The metric describes the geometry and curvature of our spacetime. This geometry is related to the matter in the spacetime via the Einstein equation:

$$G_{\mu\nu} = \frac{8\pi G}{c^4}T_{\mu\nu} , \tag{4.138}$$

where $G_{\mu\nu}$ is the Einstein tensor that encodes information about the geometry of the spacetime, while $T_{\mu\nu}$ is the energy-momentum tensor that encodes how matter and energy is distributed in that spacetime. So, the curvature of spacetime is caused by the presence of matter. Generally speaking, the more mass is concentrated in a given region, the greater the curvature of that region. If these massive objects were to move around, though, the curvature would have to change correspondingly as well in order to reflect the changed locations of these objects. In certain cases, the motion of the massive bodies causes the curvature of the spacetime to move outward at the speed of light and ripple, much like a wave. These ripples are known as *gravitational waves*.

Gravitational Waves and the Metric Perturbation

To see how this works, let's consider a region of spacetime very far from any matter. If we are sufficiently far from all matter, the spacetime will look like flat Minkowski space, and thus the line element ds of the spacetime in this region will look like:

$$ds^2 = -c^2 dt^2 + dx^2 + dy^2 + dz^2 . \tag{4.139}$$

We can denote the metric tensor for Minkowski space by $\eta_{\mu\nu}$, so the metric in this far region is given by:

$$g_{\mu\nu} = \eta_{\mu\nu} \equiv \begin{pmatrix} -1 & 0 & 0 & 0 \\ 0 & 1 & 0 & 0 \\ 0 & 0 & 1 & 0 \\ 0 & 0 & 0 & 1 \end{pmatrix} . \tag{4.140}$$

This flat spacetime has no curvature, so the Einstein tensor $G_{\mu\nu}$ vanishes. The Einstein equations are just:

$$G_{\mu\nu} = T_{\mu\nu} = 0 . \tag{4.141}$$

The components of the energy-momentum tensor are related to the local energy density and pressure flow at a given point, so there is no energy or pressure anywhere.

More realistically, we can consider a point in space that is far enough away from any matter that the gravitational effects are small but non-zero. Let's consider a spacetime that is *almost* flat Minkowski spacetime, but includes a small perturbation away from the flat metric. We can write this new metric as:

$$g_{\mu\nu} = \eta_{\mu\nu} + h_{\mu\nu} \,, \tag{4.142}$$

where $h_{\mu\nu}$ is the *metric perturbation* or *metric fluctuation*. "Smallness" in this situation is the condition that the metric fluctuation are very small:

$$|h_{\mu\nu}| \ll 1 \,. \tag{4.143}$$

Since the metric fluctuation is very small, we can do a *linearized approximation* and neglect any terms that have two or more powers of $h_{\mu\nu}$—essentially, Taylor expand around $g_{\mu\nu} = \eta_{\mu\nu}$ to first order. If we then compute the Einstein tensor for the metric $g_{\mu\nu} = \eta_{\mu\nu} + h_{\mu\nu}$, we find that the Einstein equation becomes:

$$\partial_\rho \partial^\rho h_{\mu\nu} = -\frac{16\pi G}{c^4} T_{\mu\nu} \,. \tag{4.144}$$

This is what is known as a *wave equation*: a second-order differential equation that tells us that the metric perturbation $h_{\mu\nu}$ behaves like a *wave*. (See also Sect. 3.5.2.) That is, it travels at a certain speed, with the size (e.g. the strength) of the wave oscillating both in time and space, much like waves in the ocean. The wave equation can be re-written more explicitly as:

$$\left(-\frac{1}{c^2}\frac{\partial^2}{\partial t^2} + \frac{\partial^2}{\partial x^2} + \frac{\partial^2}{\partial y^2} + \frac{\partial^2}{\partial z^2} \right) h_{\mu\nu} = -\frac{16\pi G}{c^4} T_{\mu\nu} \,. \tag{4.145}$$

Importantly, the factor of $1/c^2$ in front of the time-derivative in the wave equation indicates that the wave travels at the speed of light! This is good, because it means that gravitational signals can't travel faster than the speed of light (consistent with special relativity). We have therefore found that general relativity allows for *gravitational waves*, small ripples in the metric that propagate as waves at the speed of light.

The energy-momentum tensor $T_{\mu\nu}$ is on the right hand side of (4.145) and acts as the *source* for the small metric perturbation $h_{\mu\nu}$. However, it is not enough to have a stationary energy source to create gravitational waves. The gravitational waves oscillate in time, and correspondingly the source must also be changing in time. An example of such a system is a *binary system*, where two massive objects (such as stars or black holes) orbit around one another. The binary system will emit gravitational waves, which carry energy away from the system. Since the binary system is losing energy, the objects will lose speed until they finally come together and merge into a single object. An example of a binary black hole system emitting gravitational waves is shown in Fig. 4.13.

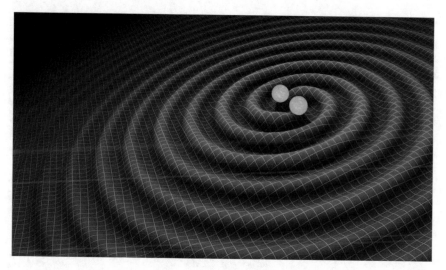

Fig. 4.13 A black hole binary system, where the black holes orbit around one another and emit gravitational waves (Image credit: R. Hurt/Caltech-JPL)

Exercise 4.56* We can define the *Weyl tensor* $C^{\mu}{}_{\nu\rho\sigma}$ as:

$$C_{\mu\nu\rho\sigma} = R_{\mu\nu\rho\sigma} + \frac{1}{2}\left(R_{\mu\sigma}g_{\nu\rho} - R_{\mu\rho}g_{\nu\sigma} + R_{\nu\rho}g_{\mu\sigma} - R_{\nu\sigma}g_{\mu\rho}\right) + \frac{1}{6}R\left(g_{\mu\rho}g_{\nu\sigma} - g_{\mu\sigma}g_{\nu\rho}\right).$$

$$(4.146)$$

Prove that $C^{\rho}{}_{\nu\rho\sigma} = 0$. Now rewrite (4.146) to see that the Riemann tensor is completely determined by the Weyl tensor and the Ricci tensor. The Ricci tensor at a given point is completely determined by the energy and momentum at that point through the Einstein equations; convince yourself that the same is not true for the Weyl tensor. In fact, the curvature contained in the Weyl tensor is due to gravitational waves—in particular, it can be non-zero at a point in spacetime even if $T_{\mu\nu} = 0$ there (i.e. if we are far away from the wave source).

Detecting Gravitational Waves

We've now established that general relativity predicts that time-varying gravitational sources (such as massive binary systems) emit gravitational waves, causing the metric to have small "ripples" on top of the Minkowski background far from the source. These arguments have been around since Einstein's original papers on relativity were published, but many scientists at the time (including Einstein himself!) were skeptical that gravitational waves actually existed.

The first experimental evidence for gravitational waves came in 1974, when astrophysicists Russell Alan Hulse and Joseph Hooton Taylor detected radio emissions from a system of binary stars, now referred to as the Hulse-Taylor binary. What they found is that, over time, the emissions from the binary actually became weaker and weaker, indicating that the binary was consistently losing energy over time.

They later showed that this loss in energy was precisely what we would expect from gravitational wave emission. This was the world's first quantitative experimental confirmation of gravitational waves, leading to Hulse and Taylor receiving the Nobel Prize in 1993.

While the Hulse-Taylor binary was a spectacular experimental success, it was only an *indirect detection* of gravitational waves. That is, the experiment tracked the inspiral of the stars and used this to lend evidence to the existence of gravitational waves, but the experiment did *not* measure the gravitational waves themselves. A long-standing goal of the physics community since then has been to measure gravitational waves *directly* and measure precisely how they interact with matter.

So, this begs the question: how do gravitational waves interact with matter? The short answer is that gravitational waves distort spacetime in such a way that the distance between objects changes. To see this, remember that the spacetime interval ds between two infinitesimally close points is given by:

$$ds^2 = g_{\mu\nu} dx^\mu dx^\nu . \tag{4.147}$$

If we now let $g_{\mu\nu} = \eta_{\mu\nu} + h_{\mu\nu}$, we find that

$$ds^2 = \eta_{\mu\nu} dx^\mu dx^\nu + h_{\mu\nu} dx^\mu dx^\nu . \tag{4.148}$$

The first term in this equation is the familiar Minkowski line element term, which is the usual spacetime interval between points in a completely flat spacetime. However, we now have a new term present due to the metric perturbation $h_{\mu\nu}$. When $h_{\mu\nu}$ is non-zero, e.g. a gravitational wave is present, the interval no longer remains the same over time! In particular, since the gravitational wave size oscillates over time, we expect to find that the distance between objects oscillates in time as well due to the passing of a gravitational wave. Two examples of how gravitational waves can change the distance between points over time are shown in Fig. 4.14.

So, in order to detect gravitational waves, we would need to find experimental evidence that shows the length of an object changing over time. This is easier said than done; gravitational waves are exceptionally weak, and they would typically make the length of an object change by only one part in 10^{20}. Luckily, we already know of one experimental apparatus that can measure lengths very precisely: the Michelson-Morley interferometer. By measuring how long light signals take to hit a mirror and then reflect back to the source, these interferometers can very accurately compare the length of the two "arms" of the interferometer.

This is the crux of how the Laser Interferometer Gravitational-Wave Observatory (LIGO) works, as pictured in Fig. 4.15. There are two metal pipes set up perpendicular to one another, each a kilometer long, with a mirror equipped at the end of each arm. Light is sent across each arm, reflected at the mirror, and then detected back at the start. By carefully timing when the light signals come back to the start, the experiment can very accurately determine how long the arms are relative to one another. If LIGO then measures that the relative difference in arms lengths changes

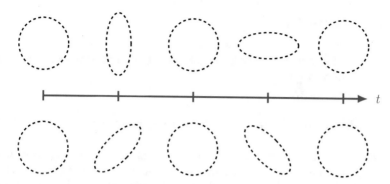

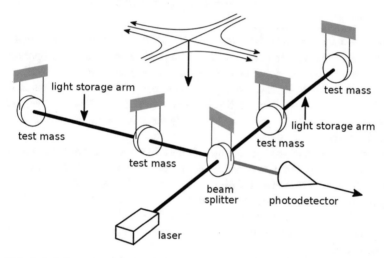

Fig. 4.14 A schematic representation of two different ways in which gravitational waves change the spatial distance between adjacent points over time

Fig. 4.15 A depiction of the LIGO experiment, which uses interferometry to precisely measure small fluctuations in lengths along the arms and therefore provide direct evidence for gravitational waves ("LIGO" by MOBle is licensed under CC BY 2.0)

periodically over time, this tells us that a gravitational wave has passed through the experiment. By making the arms one kilometer long, the experiment is incredibly precise and can track very small fluctuations in the distance. LIGO ran for over a decade; when advanced LIGO (with increased sensitivity) started its run, it almost immediately had the world's very first gravitational wave detection in September 2015. It is one of the most impressive scientific results of the past few years; the measurement of gravitational waves stands as a triumphant confirmation of general relativity. Kip Thorne, Rainer Weiss, and Barry Barish were awarded the 2017 Nobel prize in physics for the monumental work that went into this achievement.

4.6.6 Time Machines

Earlier, we talked about wormholes mainly as shortcuts through *space*, but what about shortcuts through *time*?

Consider the metric:

$$ds^2 = -(c\,dt + n\,d\phi)^2 + dr^2 + r^2 d\theta^2 + r^2 \sin^2\theta\,d\phi^2. \qquad (4.149)$$

This would be flat space if $n = 0$. When $n \neq 0$, something funny happens when $r\sin\theta < n$ (assuming n is positive). Then, we see that we can have a particle travelling along a path where $t = t_0, r = r_0, \theta = \theta_0$ are held constant and $\phi = k\tau$, because then:

$$-c^2 d\tau^2 = -(n^2 - r_0^2 \sin^2\theta_0)d\phi^2, \qquad (4.150)$$

so with $k = c/\sqrt{n^2 - r_0^2 \sin^2\theta_0}$, we have a particle travelling along a direction ϕ; the direction ϕ is a purely timelike direction. The particle feels its own proper time go forward as τ increases. However, at some point ϕ will have increased by 2π, which will take the particle back to *exactly* the same point in spacetime. This is what's called a *closed timelike curve*: a timelike curve (that a particle can travel on) that takes us from a point in spacetime to *exactly the same point* in spacetime further down the curve.

As far as time machines go, the one we described above is not quite interesting: in a sense it simply "freezes time" for the particle. However, we can also imagine altering the path of the particle slightly so that, in addition to moving "around" in ϕ, the particle also moves *back* in the coordinate t—this will still be a timelike path. The particle moves around in the ϕ direction and moves *back* in time t (even though it experiences its own proper time τ moving forward, of course): *now* we have a real time machine!

Time machines such as the one we just described violate a very important fundament of physics, i.e. that of *causality*. We mentioned the principle of causality before when we said that only *timelike-separated* points can influence each other physically. However, if we have *closed timelike loops* (i.e. time machines), all kinds of paradoxes can occur. The most popular of these is called the *grandfather paradox*, which is the subject of many science-fiction stories.

To prevent such paradoxes associated with time machines from occurring, we typically don't allow any closed time-like curves. The presence of closed timelike curves in a metric signifies that we have done something very wrong, and have not arrived at a physically meaningful metric as a result.

4.7 Beyond General Relativity

General relativity, in a sense, is the theory of *very large* numbers of particles: gravity is typically so weak that to have any interesting, measurable effect, we need a lot of particles with a lot of mass. This typically only happens when we are also considering very large distances—for example, the movement of planets around stars, or at even larger distances in cosmology. General relativity has been very successful in describing these systems of large number of particles at large distances.

On the other hand, physicists have also been very successful at understanding the interactions of a small number of particles at short distances. In this case, we turn to *quantum* mechanics, where everything exists in small packets of energy called quanta. Quantum mechanics is formulated in Newtonian space, where time is a universal parameter. We can also formulate a quantum theory on a (fixed) relativistic, flat spacetime; this is called *quantum field theory*. Note that we can ignore the curvature of spacetime that these small numbers of particles produce because gravity is so weak, or at least weak when compared to the other forces quantum field theory describes. The quantum field theory called *the standard model* has been incredibly successful in describing the many experiments done at particle accelerators. Incidentally, the reason why quantum mechanical effects are much less important when we have a large number of particles is related to a phenomenon called *decoherence*; essentially, most quantum mechanical systems can be shown to have classical behavior for large numbers of particles. We can think of the quantum fluctuations "cancelling out" in such large numbers.

Quantum field theory and general relativity are both very successful in their own respective arenas: a small number of particles at short distances, or a large number of particles at large distances. Usually, there is not really any overlap between these two arenas, so we don't need to use both theories together in any way. However, there are a few special cases where this is not true. The first example is in the center of black holes: here, we have a *large* amount of matter at *very short* distances. There is enough matter that we cannot ignore the spacetime curvature, but at the same time we are at such short distances that we need a *quantum* theory to describe the interaction of matter. Just quantum field theory or just general relativity cannot be enough to describe the center of black holes. A second example where just quantum field theory or relativity separately won't cut it is in certain aspects of cosmology. For example, there is the so-called *cosmological constant problem*: the size of the cosmological constant that simple arguments in quantum field theory would seem to predict is a whopping 10^{120} times larger than the (rather small) observed cosmological constant Λ. Since the cosmological constant can roughly be thought of as a measure of the vacuum energy in quantum field theory, this is telling us that quantum field theory is not giving us good predictions for the vacuum energy.

All this leads to the question: is there a *quantum* theory of *gravity*? The search for a consistent theory of *quantum gravity* has dominated theoretical physics research in the last few decades, and is largely still an open problem. It turns out to be exceptionally hard to marry the concept of dynamical, curved spacetime to the quantized, probabilistic nature of quantum field theory. Such a theory of quantum gravity would simultaneously account for gravitational, relativistic, and quantum phenomena, as depicted schematically in Fig. 4.16.

There are certain fundamental differences between the way quantum field theory works and the way general relativity works that make combining the theories together in a consistent framework very difficult. One problem is that quantum field theory very often involves juggling infinite energies in processes; to make physically meaningful statements, we need to *subtract out* these infinities in well-defined ways (this process is called *renormalization* of quantum field theories). On the other hand, general relativity couples to *all forms* of energy—so even, in principle, these infinite energies. How can we make sense of these infinite energies in general relativity? Does gravity really couple to them, and if so, how do we manage that they are infinite?

A second problem is that particles interact with each other in quantum field theory by exchanging other particles. Two (negatively charged) electrons will repel because they exchange a photon, the *carrier* of the electromagnetic force. However, to find correct answers for the interaction of two electrons, quantum field theory tells us we must sum over all possible ways for two electrons to exchange photons (not just necessarily one!), all at once. Now, once again, gravity couples to all particles. This means that any interaction between any particles should include summing over all

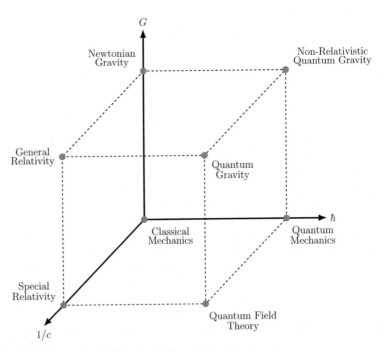

Fig. 4.16 Graphical representation of theories; moving away from the origin on one of the axes indicates taking into account that particular phenomenon. Quantum gravity must have a finite, fixed speed of light, gravity, and quantum mechanics woven into it. This cube is sometimes called the Bronstein cube, although he probably was not the first to draw it

possible exchanges of gravitons (the postulated force carrier for gravity); this essentially means summing up over all possible spacetime distortions between particles, all at once. We don't really know how to do this in a consistent way! How, then, should we make sense of gravitational interactions in a quantum field theory perspective?

One candidate—the only viable candidate so far, some will claim—of a quantum theory of gravity is *(super)string theory*. Essentially, string theory is the idea that there is no such thing as a (point) particle, but rather just one kind of string. All the particles we observe are then different excitations of the string, much like different notes on a violin string are produced by different excitations of the violin string. It is possible to quantize string theory by quantizing the string's excitations. One enormously interesting byproduct of such a quantized string theory is that it *automatically* includes gravity as one of the forces—we don't have to include it by hand, the theory *must* contain it! Thus, the quantized theory of strings automatically includes a quantized theory of gravity.

While the automatic inclusion of (quantum) gravity in string theory is a major argument in favor of it, string theory does have a lot of issues. We will only mention one problem in passing: the spacetime dimension. In (super)string theory, the number of spacetime dimensions *must be ten* for the theory to be consistent. This is obviously problematic as we appear to live in a four-dimensional world, not a ten-dimensional one. We can solve this problem somewhat by imagining the extra six dimensions are curled up into a ball so tiny that we could only really see them or the effects of their presence if we were able to produce enormously high energies, much higher than any particle accelerator we could build on Earth would ever be able to produce. However, there are a lot of ways of curling six dimensions up into a ball, which raises the question: how are they curled up in our universe? It appears that there are an enormous number of ways to do this, of the order of $\sim 10^{10^{100}}$. This is sometimes called the *landscape* of string theories. Which of these ways is the correct one that describes our universe, and why that particular one? These issues are called the *landscape problem* of string theory and their resolution is still largely not understood.

Einstein's theory of general relativity, formulated 100 years ago, must and will be improved into a consistent *quantum theory of gravity*—perhaps this will be string theory. Suffice to say that we understand so much more than we did 100 years ago, but there is still much left to explore and to understand!

More Exercises on General Relativity

Exercise 4.57** Assume there is a coordinate transformation that makes the metric *evaluated at a given point* p_0 equal to flat Minkowski spacetime $ds^2(p_0) = -dt^2 + dx^2 + dy^2 + dz^2$. At the time, this coordinate transformation makes the Christoffel symbols *at the point* p_0 equal to zero, so: $\Gamma^{\mu}_{\nu\rho}(p_0) = 0$. Note that we are simply saying that the metric and Christoffel symbols are "flat" *at a given point* p_0; this does not mean that they will be flat *everywhere*; in particular, $\partial_\sigma \Gamma^{\mu}_{\nu\rho}(p_0)$ need not be zero.

This is just like a function $f(x)$ that may have a zero $f(x_0) = 0$, without this implying that the function is everywhere zero! For a function $f(x)$ with

$f(x_0) = 0$, we can consider the Taylor expansion:

$$f(x) = f(x_0) + xf'(x_0) + \frac{x^2}{2}f''(x_0) + \cdots . \tag{4.151}$$

Any one of $f'(x_0)$, $f''(x_0)$, ... being non-zero will then imply that $f(x)$ is not zero everywhere.

The situation is similar for curved manifolds. In the coordinate transformation above, we have been able to set $ds^2(p_0)$ and $\Gamma^\mu_{\nu\rho}(p_0)$ to the "flat" values.

(a) Prove that $ds^2(p_0) = (Mink)$ and $\Gamma^\mu_{\nu\rho}(p_0) = 0$ determines the zeroth and first order derivatives of the metric $g_{\mu\nu}$ at p_0.

(b) Give a (very simplified!) expression for $R^\mu{}_{\nu\rho\sigma}(p_0)$ in terms of (derivatives of) the metric. You should get:

$$R_{\mu\nu\rho\sigma}(p_0) = \frac{1}{2}\left(\partial_\sigma\partial_\mu g_{\nu\rho} - \partial_\rho\partial_\mu g_{\nu\sigma} + \partial_\rho\partial_\nu g_{\mu\sigma} - \partial_\sigma\partial_\nu g_{\mu\rho}\right). \tag{4.152}$$

(Make sure the answer satisfies all the symmetries in Sect. 4.4.3.)

(c) Now prove that the Taylor expansion of the metric is given by (if p_0 is the origin):

$$g_{\mu\nu}(p_0) = g_{\mu\nu}(p_0) - \frac{1}{3}R_{\mu\rho\nu\sigma}(p_0)x^\rho x^\sigma + O(x^3). \tag{4.153}$$

It can help to first prove that:

$$\partial_\rho\partial_\sigma g_{\mu\nu} = -\frac{1}{3}(R_{\mu\rho\nu\sigma} + R_{\mu\sigma\nu\rho}). \tag{4.154}$$

Exercise 4.58** We have defined the covariant derivative ∇_μ. This takes a derivative along the x^μ direction. Another kind of derivative is called the *Lie derivative* $\mathcal{L}$. The Lie derivative should be thought of as taking the derivative of a tensor *along the flow of a vector field* X^μ. For example, the Lie derivative of a scalar field is:

$$\mathcal{L}_X f = X^\mu \partial_\mu f = X^\mu \nabla_\mu f. \tag{4.155}$$

The Lie derivative of a vector field Y^μ is (also defining the notation $[X, Y]^\mu$):

$$\mathcal{L}_X Y^\mu = [X, Y]^\mu = X^\nu\nabla_\nu Y^\mu - Y^\nu\nabla_\nu X^\mu. \tag{4.156}$$

(a) Prove that (4.156) is indeed a tensor by proving that you can substitute ∂_μ by ∇_μ in this expression without changing it.

(b) Using (4.155) and (4.156), derive what the Lie derivative of a covariant tensor with one index, $\mathcal{L}_X Y_\mu$, must be.

(c) What is $\mathcal{L}_X g_{\mu\nu}$? (Think of replacing all of the ∂_μ by ∇_μ to simplify the expression you get even further.) Take the metric of a sphere and calculate $\mathcal{L}_X g_{\mu\nu}$ for $X^\phi = 1$ (and other components zero). Take the Schwarzschild metric and calculate $\mathcal{L}_X g_{\mu\nu}$ for $X^t = 1$ or $X^\phi = 1$ (and other components zero). A vector X for which $\mathcal{L}_X g_{\mu\nu} = 0$ is called a *Killing vector*.

(d) A Killing vector indicates a *symmetry* of spacetime—what do we mean by this statement? Can you think of more Killing vectors for e.g. the sphere?

Exercise 4.59*** Let's derive the Schwarzschild metric. We place a mass M in spacetime (say, at the origin). This is all there is in spacetime, so we can immediately make a few observations:

- There is no preferred direction: the mass is simply a point particle (at the origin). If we place a sphere around this point particle, then no point on that sphere should be different than any other. We say the metric should be *spherically symmetric*.
- There is no time dependence: the mass will just continue to sit there (at the origin). There should be nothing that happens as t changes. We say the metric is *static*.

The unique metric that satisfies these two observations (spherically symmetric, static) is (we will not prove this):

$$ds^2 = -f(r)dt^2 + g(r)dr^2 + r^2 d\theta^2 + r^2 \sin^2\theta d\phi^2. \qquad (4.157)$$

(a) Calculate the Ricci tensor for this metric. What is the equation that the Ricci tensor must satisfy to be a solution to Einstein's equations?

(b) Use the $(\theta\theta)$ equation to express g' in terms of only f, f', g.

(c) Then use this and the (tt) equation to express g in terms of f, f'.

(d) Finally, enter the expressions you found for g, g' in the (rr) equation and find the equation governing f:

$$rf'' + 2f' = 0. \qquad (4.158)$$

(e) $f(r) = c_1/r + c_2$ is the most general solution to this equation—check that it is indeed a solution by explicitly filling it in the differential equation.

(f) Now find that $g(r) = c_2 f(r)^{-1}$ is the solution for $g(r)$.

(g) We always have a freedom of redefining a coordinate by a constant factor. Define $t' = kt$ such that the metric infinitely far away from the point particle $r \to \infty$ looks like flat, Minkowksi spacetime. Convince yourself this is essentially equivalent to setting $c_2 = c^2$.

We have now found the metric:

$$ds^2 = -\left(1 + \frac{c_1}{r}\right)(c^2 dt^2) + \frac{dr^2}{1 + \frac{c_1}{r}} + r^2 d\theta^2 + r^2 \sin^2\theta d\phi^2. \qquad (4.159)$$

The constant c_1 is the only free parameter in this metric and thus must be related to the mass of the particle sitting at $r = 0$. We can further use the weak-field approximation to relate c_1 to the actual mass of the particle. (Of course, if we use the weak-field approximation to fix the relation between c_1 and the particle's mass, we can't also use it to fix the constant of proportionality in Einstein's equations (as we did above)! Fortunately, there are other systems we can consider in the weak-field approximation to fix the proportionality constant.)

Appendix A
Quick Reference

Basic Notation

Greek letters:

α: alpha, β: beta, γ, Γ: gamma, δ, Δ: delta, ϵ: epsilon, θ, Θ: theta, κ: kappa, λ: lambda,

μ: mu, ν: nu, ρ: rho, σ: sigma, ϕ: phi, ψ, Ψ: psi, ω, Ω: omega

Einstein summation notation—only for an index that is repeated "upstairs" and "downstairs" exactly once:

$$\sum_i x_i x^i \rightarrow x_i x^i, \qquad \sum_\mu x_\mu x^\mu \rightarrow x_\mu x^\mu. \tag{A.1}$$

Useful Mathematics

Taylor expansion of function $f(x)$ around point x_0:

$$f(x) = f(x_0) + (x - x_0)f'(x_0) + \frac{(x - x_0)^2}{2} f''(x_0) + \cdots + \frac{(x - x_0)^n}{n!} f^{(n)}(x_0) + \mathcal{O}\left((x - x_0)^{n+1}\right). \tag{A.2}$$

Some useful Taylor expansions:

© Springer Nature Switzerland AG 2019
D. R. Mayerson et al., *Relativity: A Journey Through Warped Space and Time*,
https://doi.org/10.1007/978-3-030-18914-3

$$(x+a)^n = a^n + na^{n-1}x + \mathcal{O}\left(x^2\right), \quad \frac{1}{a-x} = \frac{1}{a}\left(1 + \frac{x}{a} + \frac{x^2}{a^2}\right) + \mathcal{O}\left(x^3\right),$$

$$\text{(A.3)}$$

$$e^x = 1 + x + \frac{x^2}{2} + \mathcal{O}(x^3), \tag{A.4}$$

$$\sin x = x - \frac{x^3}{6} + \mathcal{O}(x^5), \qquad \cos x = 1 - \frac{x^2}{2} + \mathcal{O}(x^4), \tag{A.5}$$

Trigonometric identities:

$$\sin(\alpha - \beta) = \cos\beta\sin\alpha - \cos\alpha\sin\beta, \quad \cos(\alpha - \beta) = \cos\alpha\cos\beta + \sin\alpha\sin\beta, \tag{A.6}$$

$$\tan 2\alpha = \frac{\cos\alpha\sin\alpha}{1 - \sin^2\alpha}, \tag{A.7}$$

$$\sin ix = i\sinh x, \qquad\qquad\qquad \cos ix = \cosh x, \tag{A.8}$$

$$\sin x = \frac{e^{ix} - e^{-ix}}{2i}, \qquad\qquad \cos x = \frac{e^{ix} + e^{-ix}}{2}, \tag{A.9}$$

$$\sinh x = \frac{e^x - e^{-x}}{2}, \qquad\qquad \cosh x = \frac{e^x + e^{-x}}{2}, \tag{A.10}$$

$$\sin^2 x + \cos^2 x = \cosh^2 x - \sinh^2 x = 1, \tag{A.11}$$

$$\text{arctanh}\, x = \frac{1}{2}\ln\left(\frac{1+x}{1-x}\right). \tag{A.12}$$

Coordinate Systems and Metrics

Cartesian coordinates $x^i = (x, y, z) = (x^1, x^2, x^3)$.
Polar coordinates in 2D:

$$x = r\cos\theta, \qquad y = r\sin\theta. \tag{A.13}$$

Spherical coordinates in 3D:

$$x = r\sin\theta\cos\phi, \qquad y = r\sin\theta\sin\phi, \qquad z = r\cos\theta. \tag{A.14}$$

Metric in Cartesian coordinates:

$$ds^2 = g_{ij}dx^i dx^j = dx^2 + dy^2 + dz^2. \tag{A.15}$$

Metric in polar coordinates (in 2D):

$$ds^2 = g_{ij}dx^i dx^j = dr^2 + r^2 d\theta^2. \tag{A.16}$$

Metric in spherical coordinates (in 3D):

$$ds^2 = g_{ij}dx^i dx^j = dr^2 + r^2 d\theta^2 + r^2 \sin^2 \theta d\phi^2. \qquad (A.17)$$

Inverse metric (with δ the identity matrix):

$$g_{ij}g^{jk} = \delta_i^k = g^{kj}g_{ji}, \qquad g_{\mu\nu}g^{\nu\rho} = \delta_\mu^\rho = g^{\rho\nu}g_{\nu\mu}. \qquad (A.18)$$

Raising and lowering indices with metric and inverse metric:

$$v_i = g_{ij}v^j, \quad v^i = g^{ij}v_j, \qquad v_\mu = g_{\mu\nu}v^\nu, \qquad v^\mu = g^{\mu\nu}v_\nu. \qquad (A.19)$$

Vector products with metric:

$$v^i w_i = v_i w^i = v^i g_{ij} w^j, \qquad v^\mu w_\mu = v_\mu w^\mu = v^\mu g_{\mu\nu} w^\nu. \qquad (A.20)$$

Vectors Products and Derivatives in 3D

Vector dot and cross product in 3 spatial dimensions:

$$\vec{v} \cdot \vec{w} = v^x w^x + v^y w^y + v^z w^z, \qquad \vec{v} \times \vec{w} = \left(v^y w^z - v^z w^y, v^z w^x - v^x w^z, v^x w^y - v^y w^x \right).$$
$$(A.21)$$

Identities:

$$\vec{A} \cdot (\vec{B} \times \vec{C}) = \vec{B} \cdot (\vec{C} \times \vec{A}) = \vec{C} \cdot (\vec{A} \times \vec{B}), \qquad (A.22)$$
$$\vec{A} \times (\vec{B} \times \vec{C}) = \vec{B}(\vec{A} \cdot \vec{C}) - \vec{C}(\vec{A} \cdot \vec{B}). \qquad (A.23)$$

Vector derivatives in 3 spatial dimensions:

$$\vec{\nabla} = \left(\frac{\partial}{\partial x}, \frac{\partial}{\partial y}, \frac{\partial}{\partial z} \right), \qquad \nabla^2 = \frac{\partial^2}{\partial x^2} + \frac{\partial^2}{\partial y^2} + \frac{\partial^2}{\partial z^2}. \qquad (A.24)$$

Basic application on vector $\vec{v}$ and function f:

$$\vec{\nabla} f = \left(\frac{\partial f}{\partial x}, \frac{\partial f}{\partial y}, \frac{\partial f}{\partial z} \right), \qquad \vec{\nabla} \cdot \vec{v} = \frac{\partial v^x}{\partial x} + \frac{\partial v^y}{\partial y} + \frac{\partial v^z}{\partial z}. \qquad (A.25)$$

Identities:

$$\vec{\nabla} \cdot (\vec{\nabla} \times \vec{A}) = 0, \qquad\qquad \vec{\nabla} \times (\vec{\nabla} f) = 0, \qquad (A.26)$$
$$\vec{\nabla} \times (\vec{\nabla} \times \vec{A}) = \vec{\nabla}(\vec{\nabla} \cdot \vec{A}) - \nabla^2 \vec{A}. \qquad (A.27)$$

Tensors and Transformations

Vector transformation rule (contravariant or "upstairs" index):

$$v'^i = \frac{\partial x'^i}{\partial x^j} v^j, \qquad v'^\mu = \frac{\partial x'^\mu}{\partial x^\nu} v^\nu. \tag{A.28}$$

Covariant or "downstairs" index transformation rule:

$$K_i' = \frac{\partial x^j}{\partial x'^i} K_j, \qquad K_\mu' = \frac{\partial x^\nu}{\partial x'^\mu} K_\nu. \tag{A.29}$$

Classical Mechanics

Position $\vec{x}$, velocity $\vec{v}$, acceleration $\vec{a}$:

$$x^i(t) = \vec{x}(t) = (x(t), y(t), z(t)), \qquad v^i(t) = \frac{d}{dt}x^i(t), \qquad a^i(t) = \frac{d^2}{dt^2}x^i(t). \tag{A.30}$$

Momentum:

$$p^i = mv^i. \tag{A.31}$$

Kinetic and potential energy:

$$E_{total} = E_{kin} + E_{pot}, \qquad E_{kin} = \frac{1}{2}mv^2, \qquad E_{pot} = V, \tag{A.32}$$

where force is given by:

$$\vec{F} = -\frac{d}{d\vec{x}}V(\vec{x}). \tag{A.33}$$

Angular momentum:

$$\vec{L} = \vec{r} \times \vec{p}. \tag{A.34}$$

Newton's laws: *total* energy *and* momentum *is always conserved without forces*. Momentum changes with a force as:

$$\vec{F} = m\vec{a} = m\frac{d}{dt}\vec{p}. \tag{A.35}$$

Different inertial observers at velocities u are related by the Galilean coordinate transformation:

$$\begin{cases} x' = x - ut, \\ y' = y. \end{cases} \tag{A.36}$$

Newton's law of gravitation:

$$\vec{F}_{grav} = -\frac{GMM'}{R^2}, \qquad G = 6.67408 \cdot 10^{-11} \frac{m^3}{kg\, s^2}. \qquad (A.37)$$

On earth:

$$\vec{F}_{grav} = -m\vec{g}, \qquad g = 9.8\frac{m}{s^2}, \qquad V_{grav} = mgy. \qquad (A.38)$$

Special Relativity

Speed of light:

$$c = 3 \cdot 10^8 \frac{m}{s}. \qquad (A.39)$$

Different inertial observers at velocities u are related by the Lorentz transformation:

$$\begin{cases} t' = \gamma\left(t - \frac{u}{c^2}x\right), \\ x' = \gamma(x - ut), \\ y' = y. \end{cases} \quad , \qquad \gamma = \sqrt{\frac{1}{1 - \frac{u^2}{c^2}}}. \qquad (A.40)$$

Spatial velocities $v = dx/dt$ then transform as:

$$v' = \frac{v - u}{1 - \frac{uv}{c^2}}. \qquad (A.41)$$

Consequences are length contraction and time dilation:

$$\Delta x' = \gamma \Delta x, \qquad \Delta t' = \gamma^{-1} \Delta t. \qquad (A.42)$$

Minkowski spacetime $x^\mu = (x^0, x^1, x^2, x^3) = (ct, x, y, z)$, with metric:

$$ds^2 = g_{\mu\nu}dx^\mu dx^\nu = -c^2 dt^2 + dx^2 + dy^2 + dz^2. \qquad (A.43)$$

Distance between two spacetime points:

$$d(p_1, p_2)^2 = -(ct_2 - ct_1)^2 + (x_2 - x_1)^2 + (y_2 - y_1)^2 + (z_2 - z_1)^2. \qquad (A.44)$$

Timelike, spacelike, null designation for separation between points and for vectors:

$$\text{Timelike:} \qquad d(p_1, p_2)^2 < 0 \qquad v^2 < 0, \qquad (A.45)$$

$$\text{Spacelike:} \qquad d(p_1, p_2)^2 > 0 \qquad v^2 > 0, \qquad (A.46)$$

$$\text{Null:} \qquad d(p_1, p_2) = 0 \qquad v^2 = 0. \qquad (A.47)$$

Proper velocity u^μ:

$$u^\mu = \left(c\gamma, \gamma v^i\right) = \frac{dx^\mu}{d\tau}, \qquad (A.48)$$

so that $u^2 = -c^2$; τ is the proper time of the particle satisfying:

$$-c^2 d\tau^2 = -c^2 dt^2 + dx^2 + dy^2 + dz^2. \qquad (A.49)$$

Momentum:

$$p^\mu = m u^\mu, \qquad (A.50)$$

with $p^2 = -m^2 c^2$, and energy:

$$E_{rel} = p^0 c, \qquad (A.51)$$

or, also valid for massless particles:

$$E^2 = \vec{p}^2 c^2 + m^2 c^4. \qquad (A.52)$$

For massless particles:

$$p^\mu = \frac{E}{c^2} (1, \vec{v}), \qquad (A.53)$$

with $\vec{v}^2 = c^2$ so that $p^2 = 0$. Conservation of energy and momentum and forces:

$$F^\mu = \frac{d}{d\tau} p^\mu. \qquad (A.54)$$

General Relativity

Principles at the basis of general relativity:
- *Einstein's equivalence principle*: All inertial (including gravity) observers are equivalent. A consequence is the *geodesic postulate*: all inertial particles travel along geodesics. Together with Mach's principle, this means that the *form* of physical laws must be the same in flat and curved spacetimes.
- *Mach's principle*: Inertia is the result of interactions between bodies. This implies spacetime must be *flat* when no matter is present, and that matter must *curve* spacetime according to where it is.

- *Principle of covariance*: The *form* of physical laws are invariant under arbitrary coordinate transformations. This implies that physical laws must be written in mathematical language using *tensors*.
- *Correspondence principle*: In the absence of gravity, physics should reduce to what we already know is correct (i.e. special relativity).

A covariant derivative ∇_μ acting on a vector gives:

$$\nabla_\mu V^\nu = \partial_\mu V^\nu + \Gamma^\mu_{\nu\rho} V^\rho, \tag{A.55}$$

where $\Gamma^\mu_{\nu\rho}$ are the connection (coefficients) or Christoffel symbols given by:

$$\Gamma^\mu_{\nu\rho} = \frac{1}{2} g^{\mu\sigma} \left(\partial_\nu g_{\rho\sigma} + \partial_\rho g_{\nu\sigma} - \partial_\sigma g_{\nu\rho} \right), \qquad \frac{\partial \vec{e}_\nu}{\partial x^\rho} = \Gamma^\mu_{\nu\rho} \vec{e}_\mu. \tag{A.56}$$

The covariant derivative transforms as a tensor; the connection itself does *not* transform as a tensor.

A geodesic $x^\mu(\tau)$ (the natural generalization of a straight line in flat spacetime) satisfies the equation:

$$\frac{d^2 x^\mu}{d\tau^2} + \Gamma^\mu_{\nu\rho} \frac{dx^\nu}{d\tau} \frac{dx^\rho}{d\tau} = 0, \tag{A.57}$$

or equivalently (with $u^\mu = dx^\mu/d\tau$):

$$u^\nu \nabla_\nu u^\mu = 0. \tag{A.58}$$

A vector V^μ is parallel transported along the x-direction if:

$$\nabla_x V^\mu = 0. \tag{A.59}$$

The Riemann tensor $R^\mu_{\;\nu\rho\sigma}$ is an indicator of curvature of the spacetime: it is completely zero if and only if spacetime is flat. Its general expression is:

$$R^\mu_{\;\nu\rho\sigma} = \partial_\rho \Gamma^\mu_{\nu\sigma} - \partial_\sigma \Gamma^\mu_{\nu\rho} + \Gamma^\mu_{\rho\lambda} \Gamma^\lambda_{\nu\sigma} - \Gamma^\mu_{\sigma\lambda} \Gamma^\lambda_{\nu\rho}. \tag{A.60}$$

From the Riemann tensor, we can form the Ricci tensor $R_{\mu\nu}$ and Ricci scalar R as contractions:

$$R_{\mu\nu} = R^\rho_{\;\mu\rho\nu}, \qquad R = g^{\mu\nu} R_{\mu\nu}, \tag{A.61}$$

and the Einstein tensor $G_{\mu\nu}$ as a linear combination of these:

$$G_{\mu\nu} = R_{\mu\nu} - \frac{1}{2} g_{\mu\nu} R. \tag{A.62}$$

The Ricci tensor and Einstein tensor are symmetric under interchange of indices (just like the metric).

On a curved spacetime, the path along which one parallel transports a vector matters. This can be written in terms of the Riemann tensor as:

$$[\nabla_\mu, \nabla_\nu] V^\rho = R_{\mu\nu\sigma}{}^\rho V^\sigma. \tag{A.63}$$

Curvature also determines how nearby geodesics converge (or diverge). For a family of geodesics $x^\mu(\tau, \lambda)$, where for fixed λ we have a geodesic, we can define $n^\mu = dx^\mu/d\lambda$; the change of n^μ as we move along nearby geodesics (i.e. increasing τ) tells us about the convergence or divergence of the family of geodesics:

$$\frac{d^2 n^\mu}{d\tau^2} + 2\Gamma^\mu_{\nu\rho} u^\nu \frac{dn^\rho}{d\tau} = u^\nu \nabla_\nu (u^\rho \nabla_\rho n^\mu) = -R^\mu{}_{\nu\rho\sigma} u^\nu u^\rho n^\sigma. \tag{A.64}$$

The energy-momentum tensor $T^{\mu\nu} = T^{\nu\mu}$ has components:

$$T^{0\mu} = T^{\mu 0} = p^\mu, \tag{A.65}$$

and satisfies conservation of energy momentum:

$$\nabla_\mu T^{\mu\nu} = 0. \tag{A.66}$$

Einstein's equations are given by:

$$G_{\mu\nu} = \frac{8\pi G}{c^4} T_{\mu\nu}. \tag{A.67}$$

Conservation of energy-momentum, $\nabla_\mu T^{\mu\nu} = 0$ is assured because $\nabla_\mu G^{\mu\nu} = 0$. Einstein's equations can be altered to include a cosmological constant Λ:

$$G_{\mu\nu} + \Lambda g_{\mu\nu} = \frac{8\pi G}{c^4} T_{\mu\nu}. \tag{A.68}$$

Some Simple Metrics

• 2D flat plane in Cartesian coordinates:

$$ds^2 = dx^2 + dy^2, \tag{A.69}$$

with $\Gamma^\mu_{\nu\rho} = 0$. The same plane in polar coordinates:

$$ds^2 = dr^2 + r^2 d\theta^2, \tag{A.70}$$

with:

$$\Gamma^r_{\theta\theta} = -r, \qquad\qquad \Gamma^\theta_{r\theta} = \Gamma^\theta_{\theta r} = \frac{1}{r}. \qquad (A.71)$$

The Riemann tensor, Ricci tensor, and Ricci scalar are all zero.
- 3D flat space in Cartesian coordinates:

$$ds^2 = dx^2 + dy^2 + dz^2, \qquad (A.72)$$

with $\Gamma^\mu_{\nu\rho} = 0$. The same space in spherical coordinates:

$$ds^2 = dr^2 + r^2 d\theta^2 + r^2 \sin^2\theta d\phi^2, \qquad (A.73)$$

with:

$$\Gamma^r_{\theta\theta} = -r, \qquad\qquad \Gamma^\theta_{r\theta} = \Gamma^\theta_{\theta r} = \Gamma^\phi_{r\phi} = \Gamma^\phi_{\phi r} = \frac{1}{r}, \qquad (A.74)$$

$$\Gamma^\theta_{\phi\phi} = -\sin\theta\cos\theta, \qquad \Gamma^\phi_{\theta\phi} = \Gamma^\phi_{\phi\theta} = \cot\theta. \qquad (A.75)$$

The Riemann tensor, Ricci tensor, and Ricci scalar are all zero.
- The 2D (curved) sphere of fixed radius R_0 has metric:

$$ds^2 = R_0^2 \left(d\theta^2 + \sin^2\theta d\phi^2 \right), \qquad (A.76)$$

with:

$$\Gamma^\theta_{\phi\phi} = -\sin\theta\cos\theta, \qquad\qquad \Gamma^\phi_{\theta\phi} = \Gamma^\phi_{\phi\theta} = \cot\theta. \qquad (A.77)$$

The Riemann tensor, Ricci tensor, and Ricci scalar are all non-vanishing:

$$R_{\theta\phi\theta\phi} = R_{\phi\theta\phi\theta} = -R_{\theta\phi\phi\theta} = -R_{\phi\theta\theta\phi} = R_0^2 \sin^2\theta, \qquad (A.78)$$

$$R_{\theta\theta} = 1, R_{\phi\phi} = \sin^2\theta, \qquad (A.79)$$

$$R = \frac{2}{R_0^2}. \qquad (A.80)$$

The Schwarzschild Metric

The Schwarzschild metric is given by:

$$ds^2 = -\left(1 - \frac{2M}{r}\right)(c^2 dt^2) + \frac{dr^2}{1 - \frac{2M}{r}} + r^2 d\theta^2 + r^2 \sin^2\theta d\phi^2. \qquad (A.81)$$

This metric describes a mass m (with $M = (G/c^2)m$) sitting at $r = 0$.
The Christoffel symbols are:

$$\Gamma^t_{tr} = \Gamma^t_{rt} = \frac{M}{r} \frac{1}{r-2M}, \quad \Gamma^r_{tt} = c^2 \frac{M}{r^3}(r-2M), \quad \Gamma^r_{rr} = -\frac{M}{r} \frac{1}{r-2M}, \quad (A.82)$$

$$\Gamma^r_{\theta\theta} = 2M - r, \quad \Gamma^r_{\phi\phi} = (2M - r)\sin^2\theta, \quad \Gamma^\theta_{r\theta} = \Gamma^\theta_{\theta r} = \Gamma^\phi_{r\phi} = \Gamma^\phi_{\phi r} = \frac{1}{r},$$
$$(A.83)$$

$$\Gamma^\phi_{\theta\phi} = \Gamma^\phi_{\phi\theta} = \cot\theta, \quad \Gamma^\theta_{\phi\phi} = -\sin\theta\cos\theta. \quad (A.84)$$

The Riemann tensor has many non-zero components, but the Ricci tensor and scalar are zero:

$$R_{\mu\nu} = R = 0. \quad (A.85)$$

The metric has a *coordinate singularity* at $r = 2M$ and a *curvature singularity* at $r = 0$. The *Kretschmann scalar* is how the curvature singularity at $r = 0$ becomes apparent:

$$R^{\mu\nu\rho\sigma} R_{\mu\nu\rho\sigma} = \frac{48M^2}{r^6}. \quad (A.86)$$

The coordinate singularity at $r = 2M$ can be removed by using the *tortoise coordinate* r^*:

$$r^* = r + 2M \ln\left|\frac{r}{2M} - 1\right|, \quad \frac{dr^*}{dr} = \left(1 - \frac{2M}{r}\right)^{-1}, \quad (A.87)$$

and going to *ingoing* (v) *or outgoing* (u) *Eddington-Finkelstein coordinates*:

$$v = t + r^*, \quad ds^2 = -\left(1 - \frac{2M}{r}\right)dv^2 + 2dv\,dr + r^2(d\theta^2 + \sin^2\theta d\phi^2);$$
$$(A.88)$$

$$u = t - r^*, \quad ds^2 = -\left(1 - \frac{2M}{r}\right)du^2 - 2du\,dr + r^2(d\theta^2 + \sin^2\theta d\phi^2).$$
$$(A.89)$$

In ingoing or outgoing Eddington-Finkelstein coordinates, there is no (coordinate) singularity anymore at $r = 2M$.

The Schwarzschild metric is the metric of a black hole. The location $r = 2M$ is the *event horizon* of the black hole; light cannot escape from within ($r < 2M$).

Appendix B
Answers to Select Problems

Chapter 2

Solution for Exercise 2.32:
You should find (approx.) $m_{\text{Earth}} = 6.24 \times 10^{24}$ kg and $R_{\text{Earth}} = 6.52 \times 10^6$ m.

Solution for Exercise 2.41(g):
For $L \neq 0$, the general solution can be written as:

$$\tilde{\theta}(t) = -\frac{gR^3}{L^2}\left(1 + C_1 \cos\left(\frac{Lt}{R^2}\right) + C_2 \sin\left(\frac{Lt}{R^2}\right)\right). \tag{B.1}$$

For $L = 0$, the general solution is quite different:

$$\tilde{\theta}(t) = \frac{g}{2R}t^2 + C_1 + C_2 t. \tag{B.2}$$

In both cases, this solution is only valid when $|\tilde{\theta}(t)| \ll 1$, which puts bounds on the range for t that this solution can be used for.

Solution for Exercise 2.42:

(a) $T = \frac{1}{2}(m_1 + m_2)\dot{x}^2$ and $V = -(m_1 - m_2)gx$ (where we ignore any constant terms in V), where x is the length of rope from the center of the disc to the mass m_1. $L = T - V$ and the equations of motion are:

$$\ddot{x} = \frac{m_1 - m_2}{m_1 + m_2}g. \tag{B.3}$$

You could have also analyzed the forces acting on each mass to obtain all the relevant equations governing their motion; the Lagrangian method is much less work to get the same answer.

© Springer Nature Switzerland AG 2019
D. R. Mayerson et al., *Relativity: A Journey Through Warped Space and Time*,
https://doi.org/10.1007/978-3-030-18914-3

(b) $T = \frac{1}{2}ml^2(\dot{\theta}^2 + \sin^2\theta\dot{\phi}^2)$ if the rod is allowed to move in all three spatial dimensions (where θ, ϕ are the spherical coordinate angles determining the rod's orientation). Note that $V = 0$.

(c) As above, we also have $V = 0$. We can split the kinetic energy T into the kinetic energy of the center of mass T_{cm} (which is at the center of the rod, i.e. fixed on the circle), and the kinetic energy about the center of mass T_{rest}. $T_{\text{rest}} = \frac{1}{2}ml^2(\dot{\theta}^2 + \sin^2\theta\dot{\phi}^2)$, just as above. $T_{\text{cm}} = ma^2\dot{\psi}^2$ where ψ is the angle along the circle. $L = T = T_{\text{cm}} + T_{\text{rest}}$.

(d) $T = \frac{1}{2}mL^2(\dot{\theta}^2 + \sin^2\theta\dot{\phi}^2)$ and $V = mgL\cos\theta$ where θ, ϕ are the spherical coordinates of the deviation of the pendulum. $L = T - V$.

Chapter 3

Solution for 3.A:

(a) In Daniel's frame, he appears to be staying still, so he will still measure the car to be of length L.

(b) $x'_F = L$

(c) The Lorentz transformations tell us that:

$$x'_F = \gamma(x_F - ut), \quad x'_B = \gamma(x_B - ut). \tag{B.4}$$

The time t here is the time at which Anthony sees Daniel's car.

(d) The length of the car that Anthony measures is $x_F - x_B$, e.g. the difference between the front and back of the car in Anthony's frame. If we take the difference between the two equations we found in part (c), we find that

$$x'_F - x'_B = \gamma(x_F - x_B), \tag{B.5}$$

where all t-dependence cancels out. We can rearrange this to write that

$$x_F - x_B = \gamma^{-1}(x'_F - x'_B). \tag{B.6}$$

And, we know from part (b) that $x'_B = 0$, while $x'_F = L$. We therefore find that the length Anthony measures is $\gamma^{-1}L$.

(e) The length contraction formula tells us that $\Delta x' = \gamma\Delta x$, where $\Delta x'$ is the length measured in the moving frame. In our problem, this is Daniel's frame, since he is moving with the car. Δx, on the other hand, is the length Anthony measures. The length contraction formula therefore tells us that

$$\text{length Anthony measures} = \Delta x = \gamma^{-1}\Delta x' = \gamma^{-1}L, \tag{B.7}$$

which is exactly what we derived in part (d). Our answer is therefore consistent with length contraction.

(f) When Anthony watches Daniel, Daniel is moving fast with respect to Anthony's frame of reference, and so Daniel's length appears to be contracted. However, motion is all relative! From Daniel's point of view, he sees himself staying still while Anthony appears to be moving. So, all of our logic described above will still apply, and so Daniel will see Anthony's length contracted. That is, Anthony will appear squished and thinner than if he were staying still in Daniel's reference frame. So, both Anthony and Daniel will see the other's length contracted, due to their relative difference in velocities.

Solution for 3.B:

(a) Daniel is traveling at a constant velocity u, and he travels for an amount of time t_F. Velocity is distance per unit time, and so the total distance he travels is ut_F.

(b) $x_F = ut_F$

(c) The Lorentz transformation formula tells us that:

$$t'_F = \gamma \left(t_F - \frac{u}{c^2} x_F \right) \tag{B.8}$$

(d) Plugging in our result $x_F = u\, t_F$ from part (b), we find that:

$$t'_F = \gamma \left(t_F - \frac{u^2}{c^2} t_F \right) = \gamma \left(1 - \frac{u^2}{c^2} \right) t_F . \tag{B.9}$$

Then, if we explicitly use the expression for the boost factor γ, we find that

$$t'_F = \sqrt{\frac{1}{1 - \frac{u^2}{c^2}}} \left(1 - \frac{u^2}{c^2} \right) t_F = \sqrt{1 - \frac{u^2}{c^2}}\, t_F = \frac{t_F}{\gamma} . \tag{B.10}$$

Rewriting this slightly, we conclude that $t_F = \gamma t'_F$. This is precisely the time dilation formula!

Solution for Exercise 3.14:

We will denote t', x' as my coordinates and t, x as your coordinates. Also, note that "y" is short for years, and "ly" for light-years.

(a) $t_1 = 1 \text{ ly}/(0.9c) = 1.111 \text{ y}$

(b) $t' = 0.484 \text{ y}$ (use the Lorentz transformation formula with $x = 1$ ly and $t = 1.1111$ y)

Solution for Exercise 3.15:

(a) In order for observer A to "see" the events, he has to actually wait for the light coming from events I and II to hit him. Since the events both occur one light-year away from him, it will take a year for observer A to see these events happen.

(b) Since observer C is travelling with velocity $-u$ with respect to A (instead of with velocity u), we need to change the sign of u in all of the Lorentz transformation formulas. The result is

$$t'' = \gamma\left(t + \frac{u}{c^2}x\right), \quad x'' = \gamma(x + ut), \quad \gamma = \sqrt{\frac{1}{1 - \frac{u^2}{c^2}}}. \tag{B.11}$$

Note that the formula for γ is unchanged, since it only depends on u^2, and so the sign of the velocity is irrelevant.

(c) See Fig. B.1 for the relevant spacetime diagram. The events occur simultaneously in observer A's frame, since they are standing still and are equally close to both events. Observer B is moving in the $+x$ direction (e.g. closer to event II), while observer C is moving in the $-x$ direction (e.g. closer to event I). Therefore event II happens before event I according to B, while event I happens before event II according to C.

We can make this more precise by using the Lorentz transformations. Recall from part (a) that observer A sees the events at times $t_I = t_{II} = 1$ y. For observer B, the Lorentz transformations tell us that

$$\begin{aligned} t'_I &= \gamma\left(1 \text{ y} - \frac{u}{c^2} \times (-1 \text{ ly})\right) = \gamma\left(1 + \frac{u}{c}\right) \times (1 \text{ y}) \\ t'_{II} &= \gamma\left(1 \text{ y} - \frac{u}{c^2} \times (+1 \text{ ly})\right) = \gamma\left(1 - \frac{u}{c}\right) \times (1 \text{ y}) \end{aligned} \tag{B.12}$$

From these expressions, we can see that $t'_I > t'_{II}$, just as predicted from the spacetime diagram. Similarly, the times for observer C are given by

$$\begin{aligned} t''_I &= \gamma\left(1 \text{ y} + \frac{u}{c^2} \times (-1 \text{ ly})\right) = \gamma\left(1 - \frac{u}{c}\right) \times (1 \text{ y}) \\ t''_{II} &= \gamma\left(1 \text{ y} + \frac{u}{c^2} \times (+1 \text{ ly})\right) = \gamma\left(1 + \frac{u}{c}\right) \times (1 \text{ y}) \end{aligned} \tag{B.13}$$

We therefore find that $t''_{II} > t''_I$, just as expected from the spacetime diagram.

Fig. B.1 Spacetime diagram for the solution to Exercise 3.15

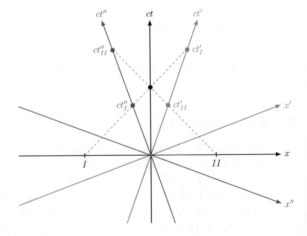

Solution for 3.C:

The resolution is that only the twin on Earth has been in an *inertial* frame the whole time. The twin in the spaceship has necessarily felt *acceleration*, most notably at the turning point of his journey (when he starts coming back). Therefore, his time dilation calculations (which are done assuming he is in an inertial frame the whole time) are actually invalid; the Earth twin's calculations are correct—he will be the older twin. (See also Exercise 3.75 where an exact calculation of the time elapsed for both twins is done.)

Solution for 3.D:

It is certainly correct to apply the length contraction formula in your reference frame: the ladder *does* indeed fit. In the ladder's reference frame, it *is* correct that the garage's length is contracted (and thus the ladder does *not* fit); the resolution to the paradox is that *in the ladder's reference frame*, the front and back doors of the garage *do not close (and immediately reopen) at the same instant in time!* This can be made quantitative using the Lorentz transformation formulae; in the ladder's reference frame, the back door (i.e. the door which the ladder travels through last) closes and reopens first, then the (front of the) ladder travels through it. Only after the (back of the) ladder has travelled through the front door of the garage, does the front door close. Thus, there is no crash; one must simply remember that the concept of simultaneity is *relative*.

Solution for 3.E:

In all three of these cases, the "thing" moving faster than the speed of light (the end of the light beam, the end of the laser, or the stick intersection point) is not actually *an object*, i.e., there is *actually no object travelling faster than light*. For example, the stick intersection point is actually two different points on each of the two sticks at every point in time; no single point on the stick is moving faster than light. No object or information can travel faster than light, and none does in any of these examples.

Solution for 3.F:

The resolution to this problem is partly similar to that of the lighthouse paradox. The cutting point of the blades is simply the same situation as in the lighthouse paradox: no object is moving faster than the speed of light since the cutting point is a different point on both the scissors and paper at each point in time. However, the ends of the scissor blades *are* one single object, so they are certainly not allowed to move faster than light. The resolution of the apparent paradox is that any physical scissors is made out of atoms and electrons which will never travel faster than light; it would theoretically require an infinite amount of energy and strength to push the scissors hard enough to make the end points of the scissor blades cut down at the speed of light. Moreover, any actual physical material that the scissors would be made out of would break into pieces long before the blades reach the speed of light.

Solution for Exercise 3.16:

When we derived length contraction, we assumed observer B (frame (t', x')) has a ruler with endpoints at x'_0 and x'_1. Implicit in this statement is that the ruler is *at rest* in B's frame, so that x'_0 and x'_1 do not change at different times (t'). Then, we can use the Lorentz transformation formulae to get $x'_1 = \gamma(x_1 - ut)$ and $x'_0 = \gamma(x_0 - ut)$, where we note that we keep the time t (now in A's frame!) fixed, since we measure x_0 and x_1 at the same time t in A's frame to determine the length in this frame. To be precise, to measure the length in A's frame, we must consider the (spacetime) distance between the two spacetime "events" $(t, x_0) \sim (t'_0, x'_0)$ and $(t, x_1) \sim (t'_1, x'_1)$. Note that $t'_0 \neq t'_1$—for $x'_1 - x'_0$ to be the length of the ruler, we rely on the fact that the ruler is at rest in the (t', x') frame!

Now, by analogy with the above, we can finally easily point out the error in the current exercise's proposed reasoning. Observer A has a ruler which is at rest in his frame with endpoints x_0 and x_1. To measure the ruler in frame B with coordinates (t', x'), we can use:

$$x'_1 = \gamma(x_1 - ut_1), \qquad x'_0 = \gamma(x_0 - ut_0), \qquad (B.14)$$

where we note that $t_0 \neq t_1$ (this is the fault in the setup for this exercise!)—in fact, we need $t'_0 = t'_1$, where:

$$t'_1 = \gamma(t_1 - u/c^2 x_1), \qquad t'_0 = \gamma(t_0 - u/c^2 x_0). \qquad (B.15)$$

Solving $t'_0 = t'_1$ gives us:

$$t_1 = t_0 - \frac{u}{c^2}(x_0 - x_1). \qquad (B.16)$$

Setting $t_0 = 0$ for simplicity leads to $x'_0 = \gamma x_0$ and $x'_1 = \gamma^{-1}(x_1 - x_0) + \gamma x_0$ so that $x'_1 - x'_0 = \gamma^{-1}(x_1 - x_0)$, as it should be.

Solution for Exercise 3.36:

(a) p_1 and p_2 are the origin at different times, and so they refer to points that are at the same spatial location as the observer at the origin. Therefore there is no spatial difference between the points, e.g $x_2 - x_1 = 0$; there is only a time difference. These points are therefore *timelike separated*. The distance between the points is

$$d(p_1, p_2) = \sqrt{-(ct_2 - ct_1)^2}. \qquad (B.17)$$

(b) If two points are timelike separated, then their separation in time is greater than their separation in space. If we choose the origin to be at one of the points, then this timelike separation means the other point must be within the first point's light cone. This is what has been shown in the first diagram in Fig. B.2. Now, let's consider boosting into a frame that is moving with a velocity u with respect to an observer sitting at the first point. If the boost velocity u is large enough, the t'-axis for the boosted observer will tilt far enough such that it actually intersects the second point, as depicted in Fig. B.2. This means that both points now occur at $x' = 0$; that is, the points now both occur at the origin in the

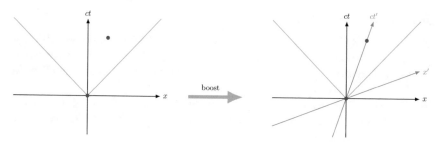

Fig. B.2 Spacetime diagram for the solution to Exercise 3.36

boosted frame, but at different times.

This means that the spatial distance between two timelike separated events depends on which frame you are in. Importantly, though, the *order* of the two events does not change! That is, for timelike separated points, all observers will agree on which event occured first.

Solution for Exercise 3.37:

(a) If you draw a spacetime diagram with p_1 and p_2 both on the x-axis, they will be spacelike separated and occur at the same time. You can also have spacelike separation without setting $t_1 = t_2$ by making sure that p_2 is outside of p_1's lightcone.

(b) To prove this, consider Fig. B.1 in the solution for Exercise 3.15 above. We start off with two spacelike separated points (I and II) with $t_1 = t_2$. Then, by boosting forwards, we find a frame in which $t_1 > t_1$. Similarly, by boosting backwards, we find a frame in which $t_1 < t_2$.

So, for any spacelike separated points, the order in which they occur is observer dependent! It does not make sense to talk about one point occuring before another, because different observers will disagree about this.

Solution for Exercise 3.39:

All observers will agree on the ordering of two events if they are *timelike* separated. This is not true for *spacelike* separated events as different observers could see the events happening in a different order; thus, the relativity of simultaneity refers to spacelike separated events.

Solution for Exercise 3.49:

(a) $\Delta t_{\text{detector}} = 2 \times 10^{-3}$ s

(b) $\Delta t_\mu = \gamma^{-1} \Delta t_{\text{detector}} = 2.82 \times 10^{-4}$ s

(c) $N(t) = 2.48 \times 10^{-39} N_0$ (it is possible to get order of magnitudes different from this answer if you used different precision for the answers in the previous parts)

Solution for Exercise 3.52:

(c) The ratio is $(1/2)(v/c)^2$, which is much less than 1 for everyday velocities, so mc^2 is always much bigger than the kinetic energy. However, for most everyday non-relativistic applications, the rest energy mc^2 does not change at all while the kinetic energy obviously does.

(d) The fraction of each uranium atom converted into (kinetic!) energy is 9×10^{-4}. For 1 g of uranium, this means a production of 8.1×10^{10} J, which can power the light bulb more than 25 years!.

Solution for Exercise 3.53:

For the non-relativistic kinetic energy, $v = \sqrt{2}c$ so $v > c$ which is not allowed by special relativity. For the relativistic kinetic energy, $v = \sqrt{3}/2c = 0.866c$ (which is safely smaller than c).

Solution for Exercise 3.67:

(a) There is no net charge in the wire, so there is no net electric force; $F_e = 0$.

(b) Use the velocity transformation formula (3.8):

$$v_\pm = \frac{v \mp u}{1 \mp vu/c^2} \tag{B.18}$$

(c) The *spacing* between successive charges will be Lorentz contracted. We have that $v_- > v_+$, so there is *more* Lorentz contraction of the space between negative charges; this means the density of the negative charge is *larger* than that of the positive charges, $\lambda_- > \lambda_+$.

If the distance between two charges *in the rest frame of the charges* is $(\Delta x)_0$, then the charge density in that frame is $\lambda_0 = (\Delta Q)/(\Delta x)_0$ where ΔQ measures the charge in $(\Delta x)_0$. In the original frame (where the negative charges are moving to the left with the same velocity that the positive charges are moving to the right), the distance between two charges (positive or negative) is $\Delta_x = \gamma^{-1}(\Delta x)_0$ (with $\gamma = \sqrt{1 - v^2/c^2}$) so that the charge density (again, both positive and negative) in the original frame is $\lambda = \gamma \lambda_0$. In the new frame, $\lambda'_\pm = \gamma_\pm \lambda_0$ where $\gamma_\pm$ are the factors associated with $v_\pm$ and can be reworked to the form:

$$\gamma_\pm = \gamma \frac{1 \mp uv/c^2}{\sqrt{1 - u^2/c^2}}. \tag{B.19}$$

The net charge density on the wire can then be found to be:

$$\lambda'_{\text{total}} = \lambda'_+ - \lambda'_- = \lambda_0(\gamma_+ - \gamma_-) = -2\lambda \frac{uv}{c^2} \frac{1}{\sqrt{1 - u^2/c^2}}. \tag{B.20}$$

(d) The electric force acting on the particle has magnitude $F'_e = qE'$, and E' is pointing *towards* the wire (since $\lambda'_{\text{total}} < 0$). If $q > 0$, the charge is *attracted* towards the wire. There is no magnetic force acting on the particle because the particle is at rest ($F_m = qv \times B = 0$).

(e) The force on the particle in the original frame is a magnetic force due to the magnetic field of the wire with current I (because in this field, the particle is moving so feels a magnetic force!). The magnetic force in the original frame is simply:

$$F_m = -quB = -qu \left(\frac{\mu_0 I}{2\pi s} \right). \tag{B.21}$$

which is the same as the force you get by transforming the electric force above, taking care that these forces are *regular* forces (i.e. not proper forces) and so transform accordingly. The relation between the regular forces in the two frames can be proven to be $F = \sqrt{1 - u^2/c^2} F'$. Also note that $c^2 = (\mu_0 \epsilon_0)^{-1}$, and $I = 2\lambda v$.

Solution for Exercise 3.70:

(a) $p^\mu = E/c(1, 1)$ and $p'^\mu = (E/c)\gamma(1 - u/c)(1, 1) = E'_\gamma/c(1, 1)$ so $E'_\gamma = \gamma(1 - u/c)E_\gamma$. It immediately follows that $f' = \gamma(1 - u/c)f$ and so $\lambda' = (\gamma(1 - u/c))^{-1}\lambda$.

(b) If $u > 0$, then $f' < f$ (and $\lambda' > \lambda$) so that the light is *redshifted*, i.e. its wavelength (or frequency) if shifted more towards the redder tints; it would be *blueshifted* if $u < 0$.

(c) If an object such as a star is moving away from us, the light that we observe from it gets redshifted. Comparing the expected frequency of light emitted from the star to the actual (redshifted) frequency of light observed by us on Earth can give us a measure of how fast the star is moving away from us.

Solution for 3.73:

This is really Exercise 3.55(c) over again in disguise. A constant proper acceleration (thus constant force) means that $a^2 = k^2$ for some real k. If we restrict ourselves to the x-plane so that $a^y = a^z = 0$, then a solution is $x(\tau) = (c^2/k)(\cosh(k/c)\tau - 1)$ and $ct(\tau) = (c^2/k) \sinh(k/c)\tau$ (if we start out at the origin). This means $x = c\sqrt{(c^2/k^2) + t^2} - c^2/k$. This is a (half) hyperbola; light rays emitted from $x = 0$ will catch up to you until $t = c/k$ (i.e. this is the head start you need); light rays emitted after that will never catch up to you!

Chapter 4

Solution for 4.A:
The text immediately following the box discusses these concepts in more detail.

(a) Here, $m = m_{inert}$, the inertial mass. A larger m_{inert} means the object "resists" a given force more (i.e. has a smaller acceleration due to the same force).
(b) Here, $m = m_{grav}$, the gravitational mass. A larger m_{grav} means gravity pulls harder on the object.
(c) They are indeed the same (or at least, always proportional with the same proportionality constant)! There is no reason for this in Newtonian physics; it is a pure coincidence!
  d $M = m_{pull}$, the amount that an object itself pulls on other objects. See also Exercise 4.1.

Solution for 4.B:
Hopefully you were able to convince yourself that *there is no experiment possible that can distinguish between situations (a) and (b)*! To distinguish (a) (and (b)) from (c), you could use e.g. two balls. For example, you could place the balls horizontally next to each other somewhere in the middle of the elevator. In situation (a) and (b), the balls would simply remain at rest (compared to the elevator); in situation (c), the balls would start moving (horizontally) towards each other, because the balls are each attracted to the point that is the *center* of the Earth. Note that you will never be able to devise an experiment that distinguishes (c) from (a) and (b) using only *one* ball (if you ignore the relative motion of the one ball compared to the elevator's walls).

Solution for Exercise 4.2:
If there is an electric or magnetic field present, we can easily detect that by having a charged particle (of which we know the charge) and comparing its motion to that of a neutral particle. In this case, we are essentially seeing the effects of two different ratios q/m. $m_{grav} = m_{inert}$ is true for *all* matter, so for any test particle we use, the ratio of these two masses is fixed—we cannot distinguish the presence of a gravitational field as we don't have any objects available with different m_{grav}/m_{inert} ratios.

Solution for Exercise 4.C:

(a) None of these concepts really make any sense; all of these properties of an object only make sense or are relevant if there are other objects to compare them to. For example, with respect to what "rest" frame would you define a velocity v of the one particle? The only possible answer is: with respect to the particle's rest frame, itself—i.e. $v = 0$!
(b) As discussed above, these concepts *must* be defined with respect to other objects. In everyday life, we could define these concepts with respect to the Earth's surface. In terms of rotation, we could also define "standing still" comparatively to the "fixed background" of stars.

Solution for 4.D:

These concepts are discussed in more detail in the text following the box.

(a) The ant has no way to *locally* "figure out" if he is on a flat sheet of paper, a curved sheet of paper, or even a ball. He could tell by certain *non-local* measurements though. For example, if the ant travelled "straight" for long enough on the ball, he would return to the spot he started out on.

(b) Two ants on parallel trajectories on a flat sheet of paper will never move closer or further away from each other—parallel lines never intersect in flat space. On the ball, the distance between the ants would change and they might even collide in the poles of the ball.

(c) A triangle on a flat sheet of paper has angles that add up to 180°. On a ball, a triangle's angles will always add up to strictly more than 180°.

Solution for 4.E:

(a) The most natural way for the ant to carry the vector around is for him to "keep it still" from his point of view. This is the parallel way to drag a vector around in Fig. 4.1.

(b) The path on a flat sheet of paper doesn't matter; "keeping the vector still" for the ant will mean the vector will always end up the same at any point on the sheet, no matter how the ant got there.

(d) Now, the path does matter! See also Fig. 4.2.

Solution for Exercise 4.12:

The first metric is simply flat 2D space in polar coordinates—thus, any vector should parallel transported along any curve in the same way that is "natural" in flat space. There are a number of ways to understand how the vector is parallel transported in the second metric. A first way is conceptual: you could realize that this metric is *locally* the same as Cartesian coordinates (x, y) (e.g. we identify r with x and θ with y); the only difference is that the θ direction is not infinite but is *compact* and periodic, where we must identify $\theta \sim \theta + 2\pi$. Thus, we can imagine taking a straight line in flat space with Cartesian coordinates, where we know how the vector is parallel transported, and "rolling" this line up into a circle. This should help visualize how the vector is transported along the circle.

A second way to understand the parallel transport in the second metric is more mathematical and uses the fact that the Christoffel symbols for this metric are all zero. Parallel transporting a vector along the circle is parallel transporting it along the θ-direction, i.e. we need that $\nabla_\theta V^\mu = 0$. Since the Christoffel symbols are zero, $\nabla_\theta = \partial_\theta$; so that we must have $\partial_\theta V^r = \partial_\theta V^\theta = 0$. In other words, the vector must look the same at each angle θ: you can imagine rotating the circle; at each point on the circle, the vector must look the same. This should also immediately help visualize the parallel transported vector.

The parallel transport of the vector for both metrics is depicted in Fig. B.3.

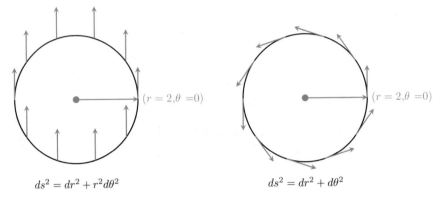

$$ds^2 = dr^2 + r^2 d\theta^2 \qquad\qquad ds^2 = dr^2 + d\theta^2$$

Fig. B.3 Illustration of parallel transport with two different metrics for the solution to Exercise 4.12

Solution for 4.F:

(a) The ant will travel on a straight line.
(b) Given the ant's position and velocity vector at that point, we have enough information to determine the next point along the ant's trajectory; this is true for any point along the trajectory.
(c) The ant's velocity vector is constant along the line: the ant carries his velocity vector and "keeps it still" from his point of view, along his own trajectory. In other words, the ant parallel transports his velocity vector along his trajectory.
(d) Being an inertial observer can also be described as having an unchanging velocity vector, *from your point of view*. In other words, the ant should again "keep his velocity vector still" (parallel transport it) along your trajectory.
(e) Once again, the next point on the ant's path is completely determined by the current point and his current velocity vector.
(f) The ant will travel on *great circles* on the sphere, as these are the geodesics of the sphere.

The key property of a geodesic is that the velocity vector of the geodesic is *parallel transported* along the geodesic. So, at every point along the geodesic, the next point is determined by the velocity vector, and the velocity vector at that next point is determined by parallel transporting itself to the new point.

Solution for Exercise 4.27:
See Fig. B.4 for an example of a negatively curved spacetime; this looks (locally) like a *saddle*. Two geodesics will *diverge*, i.e. move away from each other—the opposite of what happens on a positively curved spacetime (such as a sphere).

Fig. B.4 A negatively
curved spacetime with parts
of two geodesics on it, for
the solution to Exercise 4.27

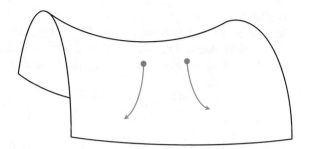

Solution for Exercise 4.36

(a) $(1/c)\partial_t \rho + \nabla p = 0$

(b) In the rest frame of the particle, $u^\mu = (1, 0, 0, 0)$, so the expression clearly gives $T^{00} = \rho$ and $T^{xx} = T^{yy} = T^{zz} = p$. Then, we notice that both sides of this expression are symmetric tensors with two (upper) indices, so the equality must be correct in any reference frame.

Solution for Exercise 4.37:

If $R_{\mu\nu} = 0$, then $R = 0$ as well. We conclude that $G_{\mu\nu} = 0$ so that Einstein's equations tell us that $T_{\mu\nu} = 0$ as well. This means that there is no matter or energy anywhere; this is indeed the case as the Schwarzschild metric is the metric of empty space outside of the matter located at $r = 0$. Only at the point $r = 0$ does $T_{\mu\nu} = 0$ not make any sense as we expect to find mass there. In fact, if we are more mathematically careful and precise, we would have found that $T_{00} \sim M\delta(r)$, i.e. the energy momentum tensor is formally infinite at $r = 0$ (due to the presence of a point mass there) and zero everywhere else.

Solution for Exercise 4.40:

(a) $v = 3.96 \cdot 10^3$ m/s; this is about $14 \cdot 10^3$ km/h.

(b) $\gamma - 1 = 8.72 \cdot 10^{-11}$.

(c) About $\Delta t = 7.53 \cdot 10^{-6}$ s, so about 7 ms. This is special relativistic *time dilation*: the moving clock (i.e. on the satellite) ticks *slower* than the one on the Earth's surface.

(d) About $\Delta t = 46.22 \cdot 10^{-6}$ s. Clocks in stronger gravitational fields tick slower, so this effect causes the clock on the satellite to tick *faster*.

(e) The special relativistic effect and the general relativistic effects are of opposite sign! The net time difference is then $\Delta t_{total} = 38.68 \cdot 10^{-6}$ s (the clock on the satellite ticks *faster*).

Solution for Exercise 4.41:

The mass of the earth is about $m = 6.24 \cdot 10^{24}$ kg and the radius is about $R = 6.52 \cdot 10^6$ m. The Schwarzschild radius $2M = 2(G/c^2)m$ is then $2M = 9.2 \cdot 10^{-3}$ m. The ratio of the aging of your head to the aging of your feet is given by:

$$\frac{d\tau_{\text{head}}}{d\tau_{\text{feet}}} = 1 + \frac{(M/R^2)}{1 - \frac{2M}{R}}\epsilon \approx 1 + 2 \cdot 10^{-16}, \tag{B.22}$$

where $\epsilon \approx 2$ m is your height. Over a lifetime (of your feet) of 100 years $\approx 3 \cdot 10^9$ s, your head will then age about:

$$(2 \cdot 10^{-16})(3 \cdot 10^9 \text{ s}) \approx 6 \cdot 10^{-7} \text{ s} \tag{B.23}$$

more than your feet. So this is an extremely tiny effect!

Solution for Exercise 4.42:

(a) An average human weighs somewhere slightly above $m \approx 50$ kg. The corresponding Schwarzschild radius is $2M = 2(G/c^2)m \approx 7 \cdot 10^{-26}$ m. This is an incredibly small size, smaller than the size of the atoms that make up our bodies! The Earth itself weighs $m = 6.24 \cdot 10^{24}$ kg, and so its Schwarzschild radius is $2M = 9.2 \cdot 10^{-3}$ m, which is far smaller than the size of the Earth itself. In both cases, the object we are talking about is much bigger than the size of its Schwarzschild radius, and so we do not have to be concerned about black holes in our daily lives.

(b) An object outside of a black hole will follow a trajectory such that, for two points infinitesimally close to one another along this trajectory, the spacetime distance ds between these points is given by

$$ds^2 = -\left(1 - \frac{2M}{r}\right)c^2 dt^2 + \frac{dr^2}{1 - \frac{2M}{r}} + r^2 d\theta^2 + r^2 \sin^2\theta \, d\phi^2, \tag{B.24}$$

where dt, dr, $d\theta$, and $d\phi$ are the differences in the spacetime coordinates (t, r, θ, ϕ) between these two points. If an object is moving along a trajectory where θ and ϕ remain constant, then any two points along the trajectory will be at the same values of θ and ϕ. We can thus set $d\theta = d\phi = 0$, leaving us with

$$ds^2 = -\left(1 - \frac{2M}{r}\right)c^2 dt^2 + \frac{dr^2}{1 - \frac{2M}{r}}. \tag{B.25}$$

(c) As discussed in Sect. 3.3.3, massive objects follow timelike trajectories in flat (Minkowski) spacetime. This is also true in curved spacetimes, like the Schwarzschild spacetime; an easy way to see this is to notice that, very far from the black hole, the Schwarzschild metric looks like the Minkowski metric (i.e. we do not notice the black hole when we are far from it), and so we better follow a timelike trajectory to match up with our knowledge of special relativity. This means that $ds^2 < 0$. For an object falling into the black hole at fixed angular coordinates, we know combine this with our answer to part (b) to find

$$ds^2 = -\left(1 - \frac{2M}{r}\right)c^2 dt^2 + \frac{dr^2}{1 - \frac{2M}{r}} < 0 . \tag{B.26}$$

When we cross into the black hole horizon, we are at a coordinate $r < 2M$, and thus $\left(1 - \frac{2M}{r}\right) < 0$. So, in the above equation, the coefficient of dt^2 becomes *positive*, while the coefficient of the dr^2 term becomes *negative*. In order to ensure that $ds^2 < 0$, then, we need to make sure that $dr \neq 0$, since the dr^2 term is the only thing that can make ds^2 negative. In order to escape the black hole after we've entered, though, our trajectory would have to have some point at which we turn around, and it is precisely at such a point that we would have a constant r along our trajectory, i.e. $dr = 0$. We are therefore not allowed to turn around once we enter a black hole!

(d) Light rays follow trajectories with $ds^2 = 0$. If we again keep θ and ϕ fixed along the trajectory, we find that

$$ds^2 = -\left(1 - \frac{2M}{r}\right)c^2 dt^2 + \frac{dr^2}{1 - \frac{2M}{r}} = 0 . \tag{B.27}$$

If we rearrange this equation somewhat and divide both sides of the equation by dt^2, we find that

$$\left(\frac{dr}{dt}\right)^2 = c^2 \left(1 - \frac{2M}{r}\right)^2 . \tag{B.28}$$

(e) As $r \to \infty$, $2M/r$ becomes very small, and so we find that $dr/dt \to c$ as $r \to \infty$. This makes sense; when we are very far from the black hole we should not notice its effects, and so the speed of the light ray should simply be its flat spacetime value, c. As $r \to 2M$, though, $2M/r \to 1$, and so we find that $dr/dt \to 0$ as the light ray approaches the black hole horizon!

(f) We see the light ray's velocity go to zero near the horizon, and so we never actually *see* the light ray cross into the black hole; it will simply keep slowing down, never quite reaching the horizon from our perspective. Another way to say this is that black holes dilate time such that the clock of an observer very far from the black hole runs infinitely faster than the clock of objects approaching a black hole horizon.

Solution for Exercise 4.48:

(a) A scalar transforms only as $S(x) \to S(x')$ under coordinate transformations. Therefore, if a scalar is infinite at a certain point in spacetime in any coordinate system, it will be infinite in *any* coordinate system. (This is not necessarily true for a vector or tensor!)

(c) The Schwarzschild metric is a solution to the vacuum Einstein equations, meaning $G_{\mu\nu} = 0$ which implies $R_{\mu\nu} = 0$. It immediately follows that $R = R_{\mu\nu}R^{\mu\nu} = 0$.

(d) Since $R_{\mu\nu\rho\sigma}R^{\mu\nu\rho\sigma}$ is a scalar (there are no free indices!), it will thus be infinite at the point $r = 0$ in any coordinate system; it is a true *curvature* singularity of the Schwarzschild solution. This singularity is precisely where the mass of the Schwarzschild solution sits. Note that none of the curvature scalars we constructed (R, $R_{\mu\nu}R^{\mu\nu}$, $R_{\mu\nu\rho\sigma}R^{\mu\nu\rho\sigma}$) show any special behavior at $r = 2M$; this is a strong indication that the singularity in the Schwarzschild metric at $r = 2M$ is a *coordinate* singularity, i.e. is an artifact of the coordinates used and not a special point in spacetime.

Solution for Exercise 4.49:

(a) If you fall into the black hole feet-first, your feet will experience a larger gravitational pull than your head. This difference in force is called the *tidal force* and is infinite at the curvature singularity at $r = 0$.

(b) When you are still outside the horizon, you will see light that is coming from behind you and falling towards the horizon, and you will also still be able to see some light that is coming from "in front of you", between you and the horizon. Just at the horizon and past the horizon, you will no longer be able to see any light from "in front of you", as that light *must* fall further into the black hole; the only light you can see inside the black hole is coming from further "outside".

(c) The external observer will never actually see you fall into the black hole. In fact, if you send out light pulses in steady intervals (according to yourself), then the observer sitting outside will see more and more time elapse between successive light pulses. The "final" light pulse that does not fall into the black hole, i.e. the one you send out right at the horizon, will actually never reach the observer outside. Of course, once you have passed the horizon, all your emitted light pulses fall further into the black hole with you.

(d) Since $r < 2M$, $2M/r - 1 > 0$. We have:

$$u^2 = c^2(\frac{2M}{r} - 1)(u^t)^2 - (\frac{2M}{r} - 1)^{-1}(u^r)^2 = -c^2, \qquad \text{(B.29)}$$

so $u^t = 0$ is certainly possible along the whole trajectory; in this case $u^r = dr/d\tau = c\sqrt{2M/r - 1}$ so that $cd\tau = dr/\sqrt{2M/r - 1}$, which we can integrate from $r = 2M$ to $r = 0$ to get $c\tau_{total} = \pi M$. If $u^t \neq 0$, then u^r must be larger than the corresponding value when $u^t = 0$; in this case we actually get $cd\tau < dr/\sqrt{2M/r - 1}$ so that $c\tau_{total} < \pi M$. So if you struggle against falling into the black hole (e.g. by turning on a rocket), you will only reach the center *faster*!

Solution for 4.G:
See also the text following the box for more discussion.

(a) When $n = 0$, the metric is simply flat (3+1)D Minkowski spacetime. l is just the usual radial coordinate and can range from 0 to $+\infty$.

(b) Now, for large $l > 0$, we again get flat (3+1)D Minkowksi spacetime.

(c) Yes! For large $l < 0$ when $n \neq 0$, the spacetime again looks like flat (3+1)D Minkowski spacetime! This is a *different* region than the flat spacetime at large $l > 0$.

(d) When $n = 0, l = 0$ is simply the origin in spherical coordinates in 3D space, i.e. it is a single *point* with 0 dimensions. When $n \neq 0, l = 0$ is a two-dimensional *sphere* of radius n.

(e) Your drawing should look something like Fig. 4.10.

(f) This spacetime is not a shortcut; the region $l > 0$ and $l < 0$ are two *distinct* regions of spacetime! However, we can imagine "gluing them together" so that they are really the same region of spacetime—see Fig. 4.11.

(g) Clearly, $G_{tt} < 0$. This implies that $T_{tt} < 0$ through Einstein's equations. T_{tt} represents the energy density of spacetime that an observer sitting at constant l, θ, ϕ measures. This does not make sense—energy densities should never be negative!

Solution for Exercise 4.55:
You should get $dR/R = \sqrt{(8\pi G\rho_0 + \Lambda)/3}\, dt$, which integrates to:

$$R(t) = R_0 \exp\left(\sqrt{\frac{8\pi G\rho_0 + \Lambda}{3}} t\right). \tag{B.30}$$

This is an *exponential* expansion (instead of the polynomial $\sim t^{2/3}$ expansion); in particular, this expansion is *accelerating* over time. By contrast, the rate of expansion for $R \sim t^{2/3}$ *slows down* (but never stops) over time.

Solution for Exercise 4.58:

(b) $\mathcal{L}_X Y_\mu = X^\nu \partial_\nu Y_\mu + Y_\nu \partial_\mu X^\nu = X^\nu \nabla_\nu Y_\mu + Y_\nu \nabla_\mu X^\nu$.

(c) $\mathcal{L}_X g_{\mu\nu} = X^\rho \partial_\rho g_{\mu\nu} + g_{\rho\nu} \partial_\mu X^\rho + g_{\mu\rho} \partial_\nu X^\rho = \nabla_\mu X_\nu + \nabla_\nu X_\mu$. For both explicit examples, $\mathcal{L}_X g_{\mu\nu} = 0$. In general, if the metric components do not depend on a particular coordinate x^i and we take $X^i = 1$ and all other components of X zero, then $\mathcal{L}_X g_{\mu\nu} = 0$.

(d) If a metric has a Killing vector or symmetry vector X, this means we can move along the X direction without changing the metric. For example, X with $X^\phi = 1$ on the sphere is a symmetry vector: moving along the ϕ direction only rotates the sphere and rotations leaves the sphere invariant (i.e. the sphere still looks the same). The sphere has three independent Killing vectors in total; you can think of these as generating rotations in each of the three spatial directions.

Bibliography

There are many books and course notes that teach special relativity, general relativity, and other related subjects. We do not know of any that aim explicitly at a level any lower than advanced undergraduate (physics major) level, but some of the undergraduate-level textbooks are quite readable without knowledge of advanced (i.e. undergraduate) mathematics or physics. We will mention a few books here that were especially inspiring to us when preparing these notes:

- J. B. Hartle, "Gravity: An Introduction to Einstein's General Relativity," Addison Wesley, San Francisco, USA (2003).
  One of the most basic books on relativity that exists. We would recommend this book for advanced students interested in expanding their knowledge of relativity beyond these notes.
- B. Schutz, "A First Course in General Relativity," Cambridge University Press, Cambridge, UK (2009).
  Assumes knowledge of but reviews special relativity; aimed at an undergraduate level with very little prerequisite knowledge of advanced physics and mathematics.
- R. D'Inverno, "Introducing Einstein's Relativity," Oxford University Press, Oxford, UK (1992).
  A good first introduction to special and general relativity at an advanced undergraduate level.
- S. M. Carroll, "Spacetime and geometry: An introduction to general relativity," Addison-Wesley, San Francisco, USA (2004).
  A modern and accessible introduction to relativity at an advanced undergraduate level.
- C. W. Misner, K. S. Thorne, J. A. Wheeler, "Gravitation," W. H. Freeman and Company, New York, USA (1973).
  The "bible" or "phone book" of general relativity. A huge, graduate-level textbook (although it does not assume much advanced mathematics), starting from the basics and working through every aspect of special and general relativity. Note that it is a little outdated in its approach to certain subjects, and it can get especially wordy in discussing certain conceptual subjects.
- R. M. Wald, "General Relativity," University of Chicago Press, Chicago, USA (1984).

© Springer Nature Switzerland AG 2019

D. R. Mayerson et al., *Relativity: A Journey Through Warped Space and Time*,

https://doi.org/10.1007/978-3-030-18914-3

A rather advanced graduate textbook in relativity that contains many advanced topics.

The following books also contain information that inspired these notes, although their primary subject matter is not relativity:

- D. J. Griffiths, "Introduction to Electrodynamics," Prentice-Hall International, Inc., Upper Saddle River, New Jersey, USA (1999).
  The quintessential advanced introduction to electrodynamics. The last chapter has a nice, brief introduction to special relativity and how electrodynamics and special relativity mesh together.
- B. Zwiebach, "A First Course in String Theory," Cambridge University Press, Cambridge, UK (2009).
  Aiming at the undergraduate level, this is certainly one of the most basic string theory books there is. It also contains some very nice sections on special and general relativity.